Data Centers Demystified

Verses Kindler Publication

Verses Kindler Publication.

Website: www.verseskindlerpublication.com

Data Centers Demystified
By: Seshu Kumar Thanneeru & Vinod Singh Yadav
ISBN: 978-81-69248-08-2
1st Edition
Price: INR 499/$22.00

Disclaimer

Data Centers Demystified by Seshu Kumar Thanneeru & Vinod Singh Yadav. The entirety of this published work comprises the original content of the author, who has undertaken diligent efforts in its editing. The characters depicted herein may be fictitious or inspired by real events; however, they are not intended to cause offense or represent any prejudice against any caste or system. Any similarity to the names of actual individuals, locations, or institutions is purely coincidental and serves the narrative. The author bears sole responsibility for the originality of this work; the publisher assumes no liability thereof.

Contents

Chapter 1

Understanding Data Centers

"There's about a trillion dollars' worth of installed base of data centers. Over the next four or five years, we'll have $2 trillion worth of data centers that will be powering software around the world." - Jensen Huang.

When you stream a movie on Netflix, transfer money using a banking app, or a doctor retrieves your medical records in an emergency, these moments are powered by facilities that most of us will never see. Much of our everyday life, whether we realize it or not, depends on something hidden from view: data centers.

Data centers are the obscure factories of the digital age: quiet, whirring buildings filled with servers and network systems that store, process, and deliver the digital content and services we use every day. They are the unseen backbone of our connected world, ensuring that the internet never sleeps and that information flows seamlessly across the globe.

But for all their importance to contemporary life, data centers still tend to be poorly understood. The term may evoke in the mind of the average person a large shed with computers and flashing lights, but a data center is far more than that. It is a sophisticated system designed to withstand and work by sustaining a delicate balance of power, cooling, connectivity, and security.

This chapter strives to provide the foundation for the data centers and their contribution to the economy, governance, sustainability, and the future of digital infrastructure. Let's understand what data centers exactly are, how they originated, and the reason they exist before we proceed with exploring policies, economics, or the underlying technology behind them.

What is a Data Center?

Essentially, a data center is an infrastructure setup that comprises servers, storage devices, computer systems and peripherals, networking appliances, and the supporting infrastructure to power them.

It is essentially the physical "backbone" of the internet and of the entire virtual world.

We recognize "the cloud" as if it were something intangible, though the cloud is really a network of data centers spread out around the globe. When you save a photo to Google Drive or open up a file from Microsoft OneDrive, that information is actually saved on a physical server or space within a data center.

Data centers serve a host of functionalities. A few primary functions that they have are:

1. Data Storage: They store enormous amounts of digital data, ranging from individual emails to business databases. In addition, they also preserve backup data to ensure that all the existing digital data remains safe from the risk of loss.

2. Data Processing: They are also equipped to continuously and rapidly process that data for more information. For example, when an online shopping website such as Amazon suggests a product to a prospective customer, algorithms on the data center servers do the computation and suggest related or similar products.

3. Data Distribution: Data centers are connected to the internet backbone in the form of a complex network. They have high-speed switches, routers, and cables, all connecting different servers to storage devices and external networks, thereby ensuring that data is delivered with speed and precision to the end users worldwide.

4. Data Security and Data Recovery: Data centers are equally committed to the safety and security of data by employing suitable data security measures. They are also equipped with measures to recover data in case it's lost due to any unforeseen circumstances.

Thus, data centers are not only meant for the storage of data; they are also dynamic centers for computation, communication, connection, and security.

The Evolution of Data Centers

Data centers emerged in parallel with the development of computing and the internet.

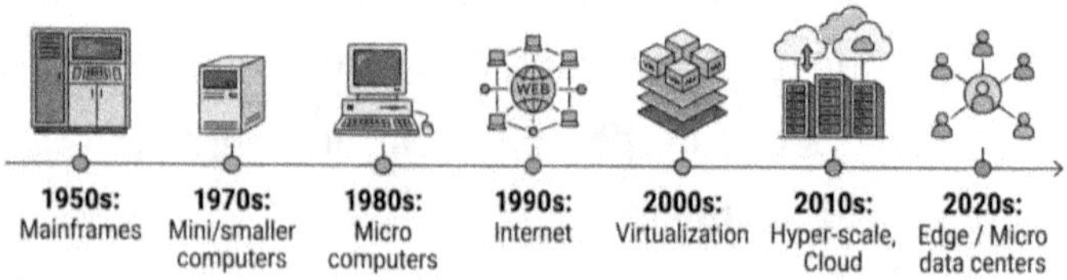

Figure 1.1 Timeline of Data Centers

1. The Early Days: Mainframes and Server Rooms

The period from the 1940s to the 1960s was a time of big mainframe concentration of computing power. Governments, universities, and large companies typically run these computers, which require special environments, air-conditioned rooms with raised floors to allow cables to pass through them, as well as to keep the equipment cool and functioning. The concept of the centralized computing room was conceived here.

In the 1970s and 1980s, with the emergence of smaller minicomputers and then personal computers, organizations began putting up their own server rooms. These were literally office building closets or small offices, crammed with rows of servers. Although these original server rooms worked, they were not necessarily secure or reliable. Cooling efficiency was frequently poor, backup systems were absent, and operational interruptions and downtimes were routine.

2. The Internet Age: Dedicated Data Centers

The 1990s saw the advent of smaller desktops and the commercial internet. Organizations had to host websites, operate email servers, and provide online services. These requirements demanded specific data centers that were purpose-built infrastructure to house computing systems and digital infrastructure. These data centers had special cooling systems, backup power, fire suppression, and physical security, unlike earlier server rooms.

The rise in the number of web hosting companies in the period also augmented the need for enterprise-level data centers. Instead of installing servers in congested office rooms, businesses were now able to rent space in facilities that had the capacity to support their applications 24/7 online.

3. Virtualization and the Cloud

The 2000s witnessed the advent of virtualization and cloud technology in a big way.

Virtualization transformed computing by allowing one physical server to operate several virtual environments in parallel, thereby making data centers more efficient, cost-effective, and capable of expanding at unprecedented speed.

Further, cloud computing arrived on the scene at about the same time. Amazon (and AWS in 2006), Microsoft, and Google started selling cloud services, enabling companies to "lease" storage, processing, and networking rather than create their own infrastructure. Behind the cloud were gigantic, complex data centers running at scales previously undreamed of.

4. Hyperscale Era

By the 2010s, the requirement for digital services burst into life with social media, e-commerce, streaming, artificial intelligence, and mobile applications all demanding gigantic computing power. The result was hyperscale data centers, which were the massive data centers built by technology giants like Amazon, Google, Microsoft, Facebook (Meta), and Apple. These centers comprised hundreds of thousands of square feet of space, filled with tens of thousands of servers, connected via super-fast networks.

Hyperscale data centers, being equipped with the latest cooling techniques, fully automated, and running on renewable energy, were more effective and robust than earlier ones.

5. The Edge Computing Movement

Although hyperscale facilities were good, the sheer volume of data generated by billions of devices, smartphones, IoT sensors, self-driving cars, and so forth created the need to process data faster and closer to the user. This gave rise to edge data centers: small and decentralized facilities that are closer to the location of data production and consumption. EDCs (Edge Data Centers), by reducing latency, made user experience seamless during real-time video streaming, autopilot driving, and online gaming.

6. Present State

Data centers form an interconnected web around the world that fuels everything from small businesses to MNCs. They vary from tiny modular facilities in small towns to sprawling hyperscale campuses in technology corridors. They are highly interlinked through a complex layout of fiber-optic networks and undersea cables.

The path from plain server rooms to advanced hyperscale facilities has been part of wider shifts in technology, business, and society. Data centers are not merely supporting infrastructures anymore; they are strategic lifelines for ordinary citizens, governments, and companies alike.

Data centers are the backbone of the digital infrastructure, and that is also what makes them the backbone of today's global digital economy. However, the present data center landscape remains varied, dynamic, and constantly evolving.

The Key Building Blocks of a Data Center

Though outwardly, a data center may look like yet another warehouse or an industrial facility, inside it is entirely a different world. Each component of the data center is carefully crafted so that operations remain uninterrupted. In most cases, data center downtime translates into millions of dollars lost per minute, like interruptions in banking and healthcare. Hence, data centers are designed with a combination of backup mechanisms, sophisticated infrastructure, and continuous monitoring systems.

The Core Elements of a Data Center: Servers and Storage

At the core of every data center lies its computing systems, primarily servers and storage devices, which form the backbone of all processing and data management activities. Typically, small- to medium-scale data centers deploy a mix of rack-mounted and blade servers designed to handle moderate workloads such as application hosting, virtualization, and database management. These servers are usually arranged in multiple racks, offering scalable performance within a compact footprint. In contrast, large-scale or hyperscale data centers operate thousands of high-density servers with multi-core processors, optimized for massive parallel computing, AI workloads, and cloud-based services. Complementing the servers are diverse storage resources, ranging from high-speed Solid-State Drives (SSDs) and traditional Hard Disk Drives (HDDs) to advanced Storage Area Networks (SANs) and Network-Attached Storage (NAS) systems, which ensure seamless access, redundancy, and data protection. Together, these computing and storage components create the powerful, resilient infrastructure that drives modern digital operations.

These core elements of the data centers are further supported by four pillars of strength: power, cooling, network, and security. As power is essential to run any digital device, likewise, a data center requires an uninterrupted power supply for its operations. To connect with and serve the outside world, these servers demand high-bandwidth networks. Cooling becomes important to preserve the systems, servers, and resources from excessive heating. Security, too, is an essential pillar to ensure that the data remains safe and free from the risks of theft or loss. All four are essential. Take away one, and down goes the entire system. Together, they are the quite vital to the virtual world.

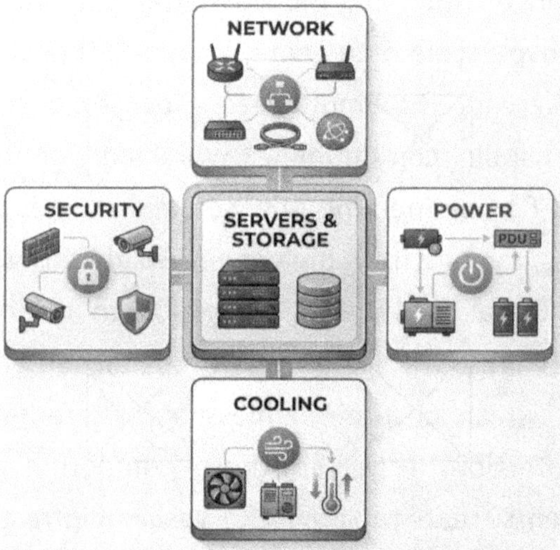

Figure 1.2 Building Blocks of Data Centers

Power: The Lifeline of Data Centers

A data center is usually referred to as an energy-hungry monster. Servers, storage arrays, and network equipment all guzzle loads of electricity. But it's not merely a matter of powering equipment; it's a matter of providing uninterrupted power, 24/7/365.

1. Primary Power Supply

Most data centers are hardwired to the electrical grid, frequently taking power from multiple substations to ensure there is no single point of failure. For hyperscale facilities, power usage can equal that of small cities. It's not uncommon for an expansive facility to use 100 megawatts (MW) or more, which is enough to light tens of thousands of houses.

Due to this tremendous demand, developers of data centers need reliable support from utility firms. Availability of power usually determines where a new data center will be erected. For instance, Northern Virginia's status as the globe's largest data center market is directly associated with its plentiful, reliable, and comparatively cheap electricity.

2. Backup Power Systems

Regardless of how solid the grid is, there are chances of breakdowns. For a data center, anything more than a few seconds of downtime is not acceptable. For instance, as per Uptime Institute's 2025 analysis on sub-minute faults, around 20% of failures cost over $1M. To guard against such breakdowns, data centers use Uninterruptible Power Supply (UPS) systems, batteries, or flywheels in banks that come into play the moment the grid fails.

For longer outages, data centers rely on diesel or natural gas generators. These generators can sustain operations for hours or even days until the primary power is restored. In most tier-rated data centers (which we'll cover later), multiple layers of redundancy are built in, meaning if one generator fails, another can immediately take over.

3. Power Distribution and Management

Electricity in a data center is not plugged directly into servers like in a home or office. Power is instead distributed via Power Distribution Units (PDUs), switchgear, and high-load circuit breakers. Each server rack can have two PDUs, powered by independent sources, so that if one fails, it can be taken over by the other.

Sophisticated monitoring systems monitor power consumption in real time, assisting operators in maximizing efficiency. PUE (Power Usage Effectiveness) is a common metric, employed as a ratio of total facility energy to the energy consumed by IT gear. An ideal PUE of 1.0 would indicate ideal efficiency (all power is consumed by IT equipment), although in practice, most data centers typically get between 1.2 and 1.6 PUE. Hyperscalers - Google and Microsoft have been leaders in methods to drive this figure lower.

4. Integration of Renewable Energy

With sustainability concerns on the rise, most operators are shifting towards renewable energy. Solar farms, wind energy, and even hydroelectric sources are being increasingly coupled with data center operations. Google, for example, has committed to operating all its data centers on carbon-free energy by 2030. Integration of renewable energy not only minimizes carbon footprints but also provides long-term energy security in areas where fossil fuel sources are unstable.

Cooling: Fending off the Heat

As data centers consist of several servers that run on electricity, they produce a lot of heat. If not adequately cooled, racks of servers can overheat and damage equipment and even cause fire risks. Cooling is thus as important as power to a data center.

Data centers use multiple systems for cooling, such as air cooling, liquid cooling, and evaporative cooling. Some of the major cooling techniques used are:

1. Conventional Air Cooling

The first data centers utilized Computer Room Air Conditioning (CRAC) units, which blew cold air into the server room. Raised floors were used to blow cool air from underneath, while hot air was blown above or through vents. While being effective, this technique requires a lot of energy/electricity.

2. Hot Aisle/Cold Aisle Containment

For enhanced efficiency, most centers currently employ hot aisle/cold aisle configurations. Servers are placed facing one another (cold aisle) while their exhausts face one another (hot aisle). This containment system physically splits hot and cold airflows apart, avoiding mixing and thus lowering the cooling requirements.

3. Liquid Cooling

As server densities are higher, air cooling alone will not be sufficient. Liquid cooling techniques, which involve liquid flow through the racks or even the processors themselves, provide an efficient alternative. Some hyperscalers have also tried immersion cooling, where servers are dipped in dielectric fluid that absorbs heat.

4. New Cooling Technologies

- Free Cooling: It involves employing external air for cooling when outdoor temperatures are low to dissipate heat.
- Seawater Cooling: Cold seawater from the deep ocean is pumped through heat exchangers to absorb heat. This technique is used in locations such as Finland.

Cooling is not just about temperature control but also about sustainability. Rather than letting heat go to waste, some centers route it to adjacent buildings or district heating networks. Stockholm, for instance, employs heat from data centers to heat homes.

Since cooling can account for nearly 40% of a data center's energy usage, innovative cooling techniques are central to both energy efficiency and environmental responsibility.

Network: Nervous System

If power is the pulse of a data center, then the network is its nervous system. Servers in isolation are purposeless without the capacity to communicate with one another and with the rest of the world.

1. Internal Networking

Within a data center, thousands of servers are connected by high-speed routers and switches. These form internal "fabrics" of communications that enable data to travel rapidly between storage, processors, and applications.

Technologies like leaf-spine architectures provide low-latency connectivity, which implies that all servers are a few hops away from any other server, thereby minimizing the delays.

2. External Connectivity

Data centers link up with Internet Exchange Points (IXPs) and telecommunications providers to connect with the external world. Some of them are also located near major fiber optic routes or landing points of submarine cables, as interconnectivity at a global level is treasured.

For instance, the city of Dublin is not only popular as a hyperscale location due to its policies but also due to its location along transatlantic fibre routes, thus providing a gateway to North America and Europe.

3. Bandwidth and Latency

The merit of a data center is usually quantified in the speed and reliability with which it can provide data. Data centers employ different bandwidths, ranging from a small bandwidth of 10 Gbps to a medium bandwidth of 100 Gbps to large bandwidths of even 1 Tbps. High bandwidth means a lot of traffic can be serviced, while low latency means high responsiveness. Providing lower-latency network connectivity brings a better experience for applications like video conferencing, share trading, or online gaming.

4. Redundancy and Resilience

Redundancy of network connectivity is essential for seamless operations. Most contemporary centers have redundant telecom carriers, varied routing circuits, and automatic failover equipment. If one network line is severed, traffic can be immediately routed through another network. The effective dynamic routing of networks and load balancing ensures that data centers remain fully functional without downtime.

Security: The Guardian Shield

Finally, security is an extremely essential factor in the context of data centers. A data center has precious data, like national secrets, financial information, or healthcare data. Intrusions into data centers, either physical or cyber, can have catastrophic effects.

1. Physical Security

Physical security is paramount to the data centers. The following are some of the security measures employed:

- Perimeter Fencing: Usually anti-ram barriers, thick concrete walls, or gabion walls.
- Surveillance: Scores of motion detection cameras, thermal imaging, or video analytics.
- Access Control: Key cards, biometric scanners, and sometimes mantraps (two-door systems that allow only one individual to enter at a given time).
- Security Staff: Trained and motivated on-site guards on 24/7 availability

Entry is strictly restricted to authorized personnel only, with multiple checkpoints.

2. Cybersecurity

The larger security threat often lies within the digital world. With the increase in cybercrimes, operators nowadays are forced to take cybersecurity as seriously as they take physical security. An information breach can cost not only huge amounts of money but also reputational loss.

Cybersecurity in data centers primarily involves:

- Firewalls and Intrusion Detection Systems (IDS)
- Encryption of data in transit as well as at rest
- DDoS protection against deluging attacks
- Real-time monitoring for unusual activity

3. Disaster Preparedness

Data centers also need protection against natural disasters like earthquakes, floods, hurricanes, and fire. Therefore, the facilities are usually constructed in low-risk areas and are equipped with seismic bracing, fire suppression, and redundant communication lines. For preparedness, mock disaster recovery drills are regularly carried out in data centers.

4. Regulatory Compliance

Security is not only about employing best practices but also about complying with relevant statutory norms and regulations. For instance, industries like finance and healthcare are subject to strict compliance with regulations such as PCI-DSS (for payment) or HIPAA (for health data in the US).

Take away

A bank cannot afford to go offline, an online store cannot afford to run at a snail's pace, and a hospital cannot afford the loss of records. A data center provides an effective solution to these concerns. The four components—power, cooling, network, and security—contribute cohesively to ensure that data centers can deliver on their promise.

Types of Data Centers

A small regional bank's IT requirements are vastly different from those of Amazon or Netflix, and thus, the data centers also vary in size, design, purpose, and ownership significantly, depending on the needs they serve.

Overall, there are different types of data centers, each having its own advantages and disadvantages, and different sets of use cases. These are on-premise data centers, colocation facilities, hyperscale data centers, edge data centers, tier-based classifications, modular centers, managed service models, and a couple of others that are evolving due to emerging technology trends.

Let's discuss each of them one by one.

1. On-Premise Data Centers

On-premise or enterprise data centers are controlled and owned by one organization for its sole utilization. They are usually found inside or close to the organization's headquarters or branch offices.

Example

A national bank can opt to have an on-site data center to safeguard sensitive financial operations and customer information in compliance with government regulations.

Characteristics

- Ownership and Control: The organization owns the entire infrastructure, land, building, servers, cooling, power systems, and networking.
- Customization: These centers are customized to the needs of the organization and typically host legacy applications or narrow workloads.
- Security and Compliance: Full physical and virtual security are under the direct control of the organization, especially in banking, healthcare, and defence organizations.

Advantages

- More control over infrastructure.
- Simpler to ensure compliance with internal or regulatory norms.
- Provide direct system access without reliance on third parties.

Challenges

- High capital outlay (CAPEX) for constructing and maintaining infrastructure.
- Limited scalability, as setting up new infrastructure takes time and resources.
- Operational complexity necessitates in-house specialists.

2. Colocation Data Centers

Colocation (colo) data centers are data centers that companies rent to house their servers and computer hardware. In this setup, the data center provider manages the building, power, cooling, connectivity, and physical security, while the company takes care of its own hardware.

Example

Smaller or mid-sized enterprises prefer to use the services of colocation companies like Equinix or NTT Global Data Centers for their operations.

Characteristics

- Customers lease racks, cages, or entire suites as per their needs.
- Fewer costs involved in shared infrastructure compared to the cost of putting up an on-premises center.

- Usually, high-speed connectivity is offered by providers.

Benefits

- Reduced CAPEX; the organizations merely pay operating costs (OPEX).
- Scalability allows firms to add more racks or cages as they grow.
- Customer avail redundant power and cooling facilities provided by operators.

Drawbacks

- Less control over the total environment as compared to the on-premises model.
- Changing provider is difficult and includes cost, once the customer gets into an agreement with the provider.

3. Hyperscale Data Centers

Hyperscale data centers are extremely large, highly automated, and specifically designed for massive data needs. They are capable of securely processing huge amounts of data and storage needs. These services are exclusively provided by tech giants like Amazon Web Services (AWS), Microsoft Azure, Google Cloud, Meta (Facebook), and Apple, and are designed to scale up to millions of users worldwide.

Example

In Ireland, Amazon, Microsoft, and Google have set up hyperscale campuses near Dublin, making it a European cloud hub. Likewise, in the United States, Northern Virginia is home to the world's most concentrated group of hyperscale centers.

Characteristics

- Require hundreds of thousands of square feet of space.
- House tens of thousands of servers.
- Usually run on renewable power and are highly efficient.
- Structured with automation and AI for monitoring and optimization.

Advantages

- Higher computational strength and exceptional scalability.
- Economies of scale lead to cost efficiency.
- Ability to venture into advanced technologies like AI, machine learning, and big data analytics.

Challenges

- Need a huge investment (billions of dollars) for setting up.
- Requires enormous power and cooling resources.
- Difficult to manage the environmental footprint considering sustainability requirements.

4. Edge Data Centers

Expansion of Internet of Things (IoT), real-time applications, and latency-tolerant services has led to the emergence of edge data centers. Unlike the hyperscale facilities, which have a centralized approach, edge centers are smaller and located closer to end users.

Example

Telecom operators deploying 5G networks are making investments in edge data centers to underpin services such as smart city applications, autonomous vehicles, and ultra-low-latency video streaming.

Characteristics

- Placed in regional towns, close to factories, telecom towers, or even at the foot of 5G cell sites.
- Offer processing power locally, lowering latency.
- Generally, supports a few hundred to thousand servers.

Advantages

- Lower latency is essential for use cases such as autonomous vehicles, online gaming, or AR/VR.
- Maximize bandwidth by first processing data locally before sending it to the cloud.
- More resilient, as local centers can keep running even if the central hub is interrupted.

Challenges

- Higher cost per unit of computation.
- Usually located in smaller towns, need higher bandwidth to connect with large cloud or hyperscale systems.

5. Modular Data Centers

Modular data centers offer a more versatile, scalable option to conventional facilities. Rather than constructing a large building brick by brick, operators employ prefabricated modules or containerized units, which can be quickly installed.

Example

Vendors such as Huawei and Schneider Electric provide modular data center solutions, commonly deployed by developing nations for disaster recovery.

Characteristics

- Built quickly with prefabricated building blocks of racks, cooling, and other power infrastructure.
- Can be used in remote locations, construction sites, or military bases.
- Modules can be removed or added, as per requirement.

Strengths

- Quick deployment possible, often within weeks.
- Scalability and flexibility.
- Low initial investment in comparison to permanent facilities.

Weaknesses

- Have limited capacity in comparison to hyperscale centers.
- Often regarded as temporary or supplementary solutions.

6. Managed Services and Cloud Data Centers

Not all organizations desire to have control over their IT infrastructure due to various reasons/limitations. Managed services providers (MSPs) and cloud data centers provide an alternative, where the provider not only hosts the infrastructure but also operates the entire setup including maintenance and compliances.

Example

AWS, Microsoft Azure, and Google Cloud are the largest players in cloud and managed data center services. Startups to multinationals utilize these platforms to host everything from websites to artificial intelligence workloads.

Characteristics

- The entire IT infrastructure is provided by the operator.
- Customers concentrate on using services rather than operating the infrastructure.
- Scalable through cloud computing paradigms, on-demand.

Benefits

- Low IT overheads for customers.
- Enterprise-class facilities available on lease.

- Pay-as-you-go pricing model minimizes initial expenditure.

Drawbacks

- Efficiency is dependent on SLAs and provider capability.
- Less customized to the customer's business requirements.

7. Tier-Based Classification

The Uptime Institute has created a worldwide accepted classification (Tiers) for benchmarking data centers based on reliability and redundancy. These "tiers" are classified on the basis of their resilience.

1. Tier I: Fundamental capacity. Limited redundancy against failures. Normally provides 99.671% uptime (~28.8 hours annual downtime).
2. Tier II: Redundant capacity. Provides 99.741% uptime (~22 hours annual downtime).
3. Tier III: Simultaneously maintainable. Devices can be replaced without a shutdown. Uptime of 99.982% (~1.6 hours annual downtime).
4. Tier IV: Fault-tolerant. Configured for uninterrupted availability, with 99.995% uptime (~26 minutes annual downtime).

Organizations choose tiers according to their business requirements and risk appetite. For example, a small company may run within a Tier II facility; however, a multinational bank would require a Tier IV facility.

8. Other New Models

The data center sector is continuously evolving, and new models are emerging progressively to meet advanced demands. Some of these models are elaborated below: -

- Micro Data Centers: Highly miniaturized facilities, typically hosting IoT or smart city applications.
- High-Performance Computing (HPC) Centers: Constructed with the capacity to manage climate modelling, scientific study, and other compute-intensive tasks.
- Sustainable Data Centers: operate on renewable power; are carbon-neutral and have a low environmental footprint.

Why So Many Types?

Data center diversity stems from the fact that companies and individuals have unique and different needs. A chain store might find it cost-effective to collocate, a government department might need on-premises management, but a video-streaming behemoth like Netflix has to depend on hyperscale facilities to support millions of users simultaneously.

Data Centers Forging Industry Change

Data centers are not mere warehouses of hardware; they are agents of change for industries. From transforming the way financial transactions happen, to fueling global health systems, to supporting e-commerce and logistics on a scale hitherto unimaginable, data centers have insidiously become central to contemporary business.

Let us take a look at some of the sectors where the difference is most evident.

Banking and Financial Services

The banking industry has been a pioneer in using information technology. Data centers have been instrumental in enabling applications ranging from mobile banking to international financial market transactions.

- Real-Time Transactions: Visa and Mastercard rely on low-latency data centers capable of supporting tens of thousands of transactions per second. A delay of just a few seconds would hamper commerce worth billions.

- Risk Management and Analytics: Banks operate sophisticated algorithms for credit scoring, fraud detection, and market analytics, possible due to data centers processing enormous data sets 24/7.

- 24/7 Global Operations: The world's stock markets open and close at different times, and data centers operate 24/7 to accommodate trading, clearing, and settlements of global stocks.
- Data Security and Compliance: Banks store highly sensitive corporate as well as personal data. Data centers employ multi-layered security, encryption, and other mechanism in compliance with standards like PCI-DSS and GDPR.

An example of a payment infrastructure that has transformed the economy of the country, is the Unified Payments Interface (UPI) in India. UPI is broadly based on extensive data center infrastructure, capable of handling billions of transactions every month.

Healthcare

Data centers are contributing a lot to the healthcare industry. The contemporary healthcare sector produces vast quantities of data, electronic health records (EHRs), diagnostic images, genomic sequences, and clinical trial data, all of which must be stored, processed, and exchanged securely.

- Electronic Health Records: Centralized data centers are what hospitals and clinics use to stock patient records. This enables diversely located physicians to readily access medical histories of patients, enhancing diagnosis and care. The records are maintained securely, in complete compliance with the global HIPAA regulations.

- Telemedicine: COVID-19 pandemic accelerated the acceptance of telehealth worldwide. Edge data centers have played a significant role in enhancing telemedicine by processing large amounts of data locally to offer low-latency communication with quick response times, ensuring seamless video consultations between physicians and patients, while also providing robust security to sensitive health records.
- Medical Imaging and AI Diagnosis: Medical scans, such as MRIs or CTs, generate massive amounts of data. Data centers enable the processing of this data and execute AI programs to identify anomalies, in some cases, swifter than human radiologists.
- Genomics and Research: Cracking the human genome or studying pandemic-related mutations require high-performance computing (HPC) centers, capable of handling high workload.

An excellent example would be how, in the COVID pandemic, data centers created on the cloud led to international coordination in the development of vaccines.

Scientists around the world accessed common datasets stored on secure, scalable infrastructure, reducing vaccine development time significantly.

E-Commerce and Digital Services

The e-commerce sector is arguably the most visible user of data centers. Online Shopping sites, tracking through logistics, and targeted advertising all rely on robust infrastructure.

- Platform Availability: Retail giants Amazon, Flipkart, and Alibaba cannot afford any downtime during holiday seasons. Their hyperscale data centers support uninterrupted availability even with massive traffic volumes.
- Recommendation Engines: Data centers drive algorithms that recommend products based on browsing behavior and previous purchases.
- Inventory and Logistics: Supply chain management systems with real-time inventory tracking are hosted in data centers, enabling sellers to monitor inventory levels and logistics partners to track shipments.
- Payment Gateways: data center support compliant payment processing systems through strong encryption and other security techniques.

The unprecedented online growth during the pandemic underscored the strength of data centers. As physical stores closed, online stores prospered due to the infrastructure that scaled on demand to address the record demand.

Other Industries

1. Media & Entertainment: Streaming services like Netflix, YouTube, and Disney+ use content delivery networks (CDNs) through distributed data centers spread across the world. Without the local caching available in data centers, buffering would make streaming unwatchable.

2. Education: Universities are increasingly using Online learning centers, supported by cloud data centers to offer education to students globally.

3. Manufacturing & IoT: Edge data centers capable of effectively analyzing sensor data on the shop floor, are making industrial automation more efficient and secure.

4. Government Services: Digital governance initiatives like Aadhaar and Smart cities, all rely on secure and scalable data centers.

A Sneak Peek: Data Centers and Economic Growth

As we will explore in forthcoming chapters, data centers are not only technical facilities but are also economic drivers. Construction and maintenance of large facilities like data centers entail billions of dollars of investment, thousands of employment opportunities, and has great spillover effects on energy, telecom, logistics, and hospitality industries.

As ports and airports were considered strategic assets in the industrial era, similarly data centers are increasingly becoming crucial in the digital era. Economic stakes are high, and industrial policies will determine national growth trajectories. Data localization policies, tax incentives, and sustainability mandates will define the countries that become hubs of the world. This is why governments across Singapore to India, Ireland to the United States is competing to entice hyperscale operators.

Case Study 1: Singapore - From IT Parks to Global Hubs

Singapore today is one of the top data-center markets in Asia-Pacific both in terms of capacity (1,000 MW) and value (USD 3–4+ billion in 2025), and is regarded as the preferred destination by hyperscalers and global financial institutions.

- Early Beginnings: In the 1980s and 1990s, Singapore started to develop government-supported IT parks to woo multinational companies. Singapore offered the best infrastructure, tax incentives, and competent workforce to entice the companies.
- Strategic Location: Located at the intersection of Asia, Singapore was a natural choice for connectivity with

submarine cables connecting it to the world's leading markets.

- Policy Support: The government actively fostered data center development with incentives, simplified approvals, and investments in infrastructure. The Infocomm Development Authority (IDA) played a central role in the policy formulation.
- Global Players: Today, global giants like Amazon, Microsoft, Google, Facebook, and others have their hyperscale facilities in Singapore. Thus, the city-state has one of the highest data center densities per square kilometer in the world.
- Challenges: Accelerated growth also presented its set of challenges. Power consumption and sustainability issues prompted a brief moratorium on new construction of data centers in 2019. This phase, however, was utilized to embrace green energy and better sustainability practices.

Singapore has shown the ideal path with timely planning, clear policymaking, and targeted investment, turning a small country into a digital infrastructure powerhouse.

Case Study 2: Dublin, Ireland: Europe's Hyperscale Hub

Dublin hosts an overwhelming majority of hyperscaler like Amazon, Google, Microsoft, Meta and others, thereby concentrating European cloud infrastructure in single metro region. It is an ideal case study depicting how a small market can become a top-tier hyperscale hub.

- Geographic Advantage: Location of Dublin on Europe's east coast, make it closest to transatlantic subsea fibre routes, enabling the shortest paths between North America and Europe.
- Business Environment: Ireland's friendly tax regime, educated workforce, and political stability have brought the investments of Amazon, Google, Microsoft, and Facebook.
- Critical Mass: With passage of time, Dublin reached a tipping point. After initial establishment of facilities by key players, an ecosystem of vendors, engineers, and connectivity providers set in, further solidifying its hub status.
- Scale and Growth: Dublin is home to dozens of hyperscale data centers, not only for Ireland but for much of Europe. Its facilities are imperative for cloud services in the European Union.
- Sustainability Issues: Just like Singapore, high growth produced sustainability issues in Dublin, with data centers consuming large amounts of the national power supply. Ireland is now confronted with the

need to balance development with green responsibilities.

Dublin shows how a mix of location, policy, and ecosystem creation can produce world-class centers.

The Foundation for Shaping a Future

This opening chapter has provided a foundational understanding of data centers, their evolution, key components, types, and transformative role across industries. We also saw how forward-looking policies and strategic positioning have enabled countries like Singapore and Ireland to become global leaders in this space.

As we enter the next chapter, we move the emphasis from the "what" and "how" of data centers to the "why" of why they are crucial in the international economy. The policies, investments, and industry linkages surrounding data centers are not just driving digital transformation but also determining the economic destinies of countries.

Chapter 2

Data Centers and Economic Growth

"Data centers are the physical backbone of the digital economy." - Centre for Economic Policy Research (CEPR)

The Economic Pulse of Data Centers

A data center, outwardly, appears like an infrastructure made up of concrete, steel, and glass designed to house servers and other IT hardware. But behind the humdrum exterior is an ecosystem driving much of the economic activity happening today. Each cable, processor, and fan within these buildings means jobs, taxes, investment, and growth that ripple far beyond the doors.

Each file uploaded, each online payment validated, each video call finished adds a slight beat to a worldwide network of digital infrastructure. Those beats, when combined through billions of interactions, have become a major driver of economic worth, usually invisible but profoundly transformational.

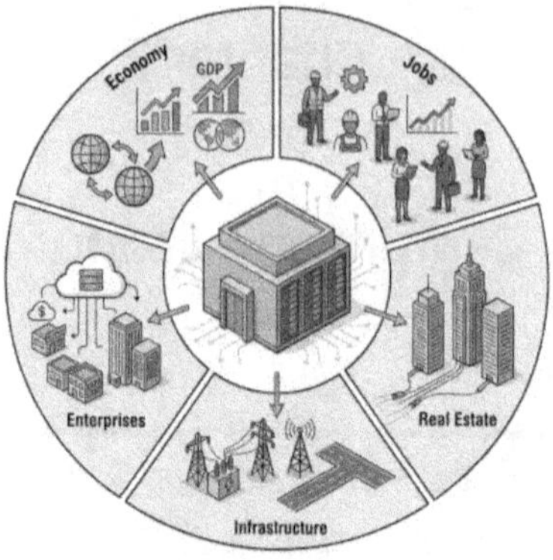

Figure 2.1 Impact of Data Centers

Data centers are not just repositories of information. They are catalysts of economic development. Their growth triggers a chain of benefits that touch nearly every part of the economic fabric:

- Growth: through investments, capital inflows, and sustained GDP contribution.
- Jobs: by creating new categories of employment from construction to high-tech operations.
- Real Estate: by revitalizing urban and semi-urban areas into digital corridors.
- Infrastructure: by accelerating upgrades in power, transport, and connectivity networks.
- Enterprises: by enabling start-ups and small businesses to compete with global agility through affordable access to cloud and computing power.

Thus, in essence, data centers are not mere back-end utilities, they are front-line enablers of progress. They transform cities into digital hubs, industries into data-driven ecosystems, and workforces into future-ready communities. Their economic heartbeat now resonates through every sector, redefining what development means in the digital age.

The Direct Economic Footprint

The most overt manner in which data centers are a part of growth is through direct economic activity, the actual building, the operations, and the expenditures that come along with both.

Construction as Capital Formation

When a hyperscale provider commits to constructing a data center, a domino effect is triggered. Engineers create high-density server configurations; steel is erected and foundations are poured by civil contractors; welders, electricians, HVAC (Heating, ventilation, and air conditioning) technicians, and network professionals follow. Every project may need hundreds of skilled as well as semi-skilled labourers over 12–24 months.

According to the Knight Frank–NASSCOM "India Data Center Report 2024," the average construction cost per megawatt of data center capacity in India currently ranges between ₹40–50 crore, depending on city and configuration.

This means that a 30 MW facility typically represents a total project investment of about ₹1,200–1,500 crore, a scale comparable to large industrial plants or power-generation units. Such projects also generate extensive multiplier effects, supporting hundreds of contractors, suppliers, and on-ground service providers in the local economy. Even the transport companies carrying servers and air-conditioning units get tangible gains.

Contrary to most other infrastructure schemes, which dwindle after opening, data centers move into a second stage of prolonged economic life when they are running.

Operational Expenditures and Ongoing Outlays

Once the ribbon is cut, a facility continues to inject funds into the neighborhood through routine operations. Power purchase deals, water consumption, maintenance agreements, security services, and IT operations all create constant revenue streams for local suppliers.

A typical hyperscale campus has between 100 and 300 full-time technical personnel, engineers, facility managers, and network administrators, and indirectly supports another 600 to 1,000 indirect positions through janitorial, catering, transport, and third-party maintenance services. As a result, the operational revenue of a data center in the USA is estimated to reach $171 billion by the end of 2025.

Electricity usage can make up 40–50 percent of the total cost of running, thereby providing reliable demand to local power utilities. Supply of cool water, waste, and supply chains around these plants creates micro-economies, little groups of specialized companies that live almost entirely on data center deals.

Tax Revenues and Fiscal Contribution

Governments also see data centers as profitable. Property taxes, electricity-duty collections, and GST (on supplies and services) come in regularly once operations are initiated. Further, import duties on costly IT hardware, racks, networking switches, and processors contribute to revenue inflows. States like Nebraska alone reported over $1.3 billion in combined tax revenues from data center operations in 2022 during new data center constructions.

In addition to revenue, the industry creates long-term investment in energy and transport infrastructure, substations, fibre corridors, and access roads, which ultimately raise a region's industrial competitiveness as a whole.

The Ripple Effect: Indirect and Induced Impacts

Whereas direct spending is the visible tip of the iceberg, indirect and induced effects are the hidden mass that underpins entire local economies.

Vendor Ecosystems and Logistics Chains

Each data center depends on a complex network of suppliers, power-equipment companies, cable makers, security-system installers, and software integrators. When a cluster of these buildings grows up, vendors start moving in nearby, cutting lead times and logistics expenses. This localization generates industrial clusters akin to those formerly in existence around car factories or oil refineries.

For instance, in Navi Mumbai, where there are several dozen facilities concentrated, logistics firms now have specialized fleets for server carriage under sealed, humidity and shock-free conditions. High-tech construction companies that used to cater to petrochemical plants diversified into data center fit-outs, working out raised-floor systems and liquid-cooling infrastructure. The domino effect is huge: whenever one sector improves, others adapt to its standards.

Hospitality and Service Spillovers

Technicians, auditors, and global teams traveling to commission or maintain these facilities also drive the hospitality and service economy. Hotels near hyperscale campuses in Hyderabad's Financial District or Chennai's Siruseri belt often report steady occupancy from data center personnel and vendors. Catering services, housekeeping contractors, and even retail outlets benefit from the constant presence of a high-skill workforce.

Alongside, training schools, safety equipment manufacturers, and certifying bodies emerge to cater to this infrastructure. In Pune and Noida alone, numerous small vocational institutions now teach specialized facility management and data center safety procedures courses, a field nearly non-existent a decade back.

Urban Development and Infrastructure Uplift

Major data center investments tend to push councils to improve supporting infrastructure, upgrade roads, enhance drainage, and support power networks. Upgrades, in turn, draw in unconnected businesses and residential units. Hence, one data center cluster may serve as an anchor for wider city regeneration.

For example, what was once an industrial suburb of Mahape (Navi Mumbai) has become a hybrid-use technology park. Real-estate demand accelerated not just for office parks but for housing for engineers and service professionals. Domestic property values have been said to appreciate 20–25 percent since the initial wave of facilities was commissioned.

The Multiplier in Numbers

Economists use the output multiplier to gauge the impact of spending on economic activity. In construction-heavy sectors like data centers, this multiplier ranges from 1.8 to

2.2, meaning every ₹1 spent can generate nearly ₹2 in overall economic activity.

In the U.S., large-scale data center investments have become a significant driver of GDP growth. Economist Jason Furman from Harvard notes that in the first half of 2025, spending in data centers and information processing accounted for almost all of the country's GDP growth. J.P. Morgan estimates that data center spending could increase U.S. GDP by 10-20 basis points in 2025-26. This evolution marks data centers as a crucial growth engine for the digital economy, creating high-value jobs and driving ancillary economic effects.

Data center investments are comparable to large factories in their growth potential, but they have a key advantage: digital infrastructure remains relevant longer due to ongoing operational costs and technology updates.

When calculating the employment impact of a data center, both direct and indirect jobs must be considered:

- Direct employment: 300–500 per facility
- Indirect employment: 1,500–2,000 (vendors, logistics, security, etc.)
- Induced employment: 2,000+ (retail, transportation, services)

In total, one large hyperscale facility can support around 3,000-4,000 jobs. With India's anticipated 500+ MW capacity addition each year, this translates to tens of thousands of stable, skilled, and semi-skilled employment opportunities across states. Unlike gig economy jobs, these positions are linked to physical assets, ensuring job security and career growth.

The Invisible Engine of Productivity

Though direct and indirect effects are concrete, data centers also promote systemic productivity benefits that percolate into all digital transactions in the economy.

Lower Transaction Costs Across Industries

Faster handling of financial transactions, facilitated by powerful back-end data centers, cuts down on transaction delay and increases liquidity in the market. Every millisecond saved in financial settlements adds up to collective national productivity.

Likewise, producers using Internet-of-Things (IoT) information for predictive maintenance reduce downtime and maximize the use of resources. In the absence of local nodes for data processing and cloud back-ends, real-time analytics like these would not be feasible.

Digitalization of Conventional Industries

Agriculture, logistics, health, and education all experience secondary benefits when there is a surety of computing infrastructure. Start-ups developing precision-farming equipment or telemedicine platforms can scale up because they can lease compute power rather than spend millions on hardware.

This democratization of technology, facilitated through affordable data center capacity, gives rise to broad-based economic growth as opposed to high-tech enclave growth.

Data Center and the Movement of Capital

In addition to domestic spillovers, data centers have become magnets for international capital. These are seen by foreign investors not just as real-estate holdings but as long-term infrastructure investments akin to airports or electricity grids.

Foreign Direct Investment (FDI) Attraction

Over the past half-decade, India has seen a flood of FDI into data center projects. Multinationals like Amazon Web Services (AWS), Google Cloud, Microsoft, and Equinix have made multi-billion-dollar proposals, usually in the form of joint ventures with local companies.

These investments have a number of characteristics:

1. Long gestation, stable returns – attractive to pension funds and sovereign wealth investors.
2. Technology transfer – implementation of state-of-the-art cooling, monitoring, and automation systems.

3. Skill enhancement – best practices from around the world imparted to local operators.

A notable development is the plan to establish a 1 gigawatt AI data center in Vizag, which is projected to involve an investment of $15 billion. This adds to the ongoing initiatives, such as Google's ₹33,000-crore (≈ USD 4 billion) India Digital Hub, which includes new cloud regions and accompanying training programs. Additionally, AWS has committed to a USD 12.7 billion investment in Indian data centers by 2030, anticipating an economic impact of over ₹1 lakh crore both directly and indirectly.

These commitments send powerful messages to other international investors that India's digital infrastructure is structural, not speculative.

Capital Inflows Over FDI

Domestic institutional investors have caught up, too. Infrastructure funds, real-estate investment trusts (REITs), and private-equity firms now consider data centers as an asset class in their own right. The long-term lease arrangements and stable energy demand make them suitable for portfolio diversification.

This has generated ancillary capital flows into fiber networks, renewable-energy joint ventures, and transmission lines. Every rupee put into data center properties is likely to attract another rupee of investment into supporting infrastructure.

Building the Foundations of a Digital Economy

A Network Effect for Countries

The aggregate impact of these investments, building, operations, capital flows, and real-estate conversion, produces what economists call a network effect at the level of nations. The more digital infrastructure there is, the lower the cost of digital participation for all.

As connectivity gets more affordable and cloud offerings become more dependable, small enterprises digitalize more rapidly, e-governance portals reach distant citizens, and research institutions become connected to global data sets. Essentially, data centers help to magnify every other type of investment, education, healthcare, and manufacturing by providing them with computational muscle.

The Compounding Curve

The expansion effect of the data center is non-linear. Early sites are for large corporations primarily, but as size increases and costs decrease, it expands to SMEs, start-ups, and ultimately to individual developers or creatives. This snowball effect replicates the diffusion of electricity a century earlier: once electricity was everywhere, innovation snowballed. Now, computational capacity does the same.

It is no hyperbole then to call data centers the new power stations of the digital economy, as they transform electrical energy into information energy, which in turn powers productivity and growth.

Drivers of Growth: FDI, Capital Flows, and Real-Estate Spillovers

---

"The world's information no longer comes alone; it brings its cash with it." – Karl Polanyi.

When a nation becomes a data center hub, it doesn't simply keep servers; it is a repository of capital, knowledge, and trust. Each rack deployed, each megawatt coming online, is not only evidence of technological advancement but proof of investor assertion that the area is prepared for the next phase of the digital economy.

The ancient game entailed the fight of countries over factories, ports, and mines. They now vie over data gravity, the authority to draw, store, and process data of the globe. And with that, data is something equally precious - an investment.

A Magnet for Global Capital

Foreign Direct Investment (FDI) in data centers is not merely money inflow; it is an indicator of the way the world perceives a nation's preparedness in the digital domain. International technology leaders and infrastructure funds look for places with secure power, good connectivity, and favorable regulation. When they make investments, along with money, they bring international standards, technologies, and job opportunities in the long term.

India's FDI inflows in the data center industry have increased manifold over the past decade. Announced and committed projects by hyperscalers, colocation companies, and infrastructure funds between 2018 and 2024 have reached more than USD 20 billion, with estimates indicating the market could reach USD 35–50 billion by 2030. Moreover, India, too, is aiming to build data centers with 9GW capacity by 2030.

These figures are more than economic size; they are a reflection of faith in India's capability to build large-scale digital infrastructure. Investors like Brookfield, Blackstone, Ascendas-Singbridge (CapitaLand), and Digital Realty have signed deals with Indian companies, wagering on the long-term potential of India's digital revolution.

The AWS and Google Effect

Amazon Web Services (AWS) and Google Cloud have been very important in this scramble.

AWS India: AWS has committed USD 12.7 billion to expand its cloud infrastructure in India to twofold by 2030 and focuses on its existing sites in Mumbai and Hyderabad (May 2023). The company has estimated that its activity will add ₹1.94 lakh crore (USD 23 billion) to the GDP of India between 2023 and 2030, creating 131,700 full-time equivalent jobs each year across industries.

Google Cloud: Google's India Cloud Region, which opened in Mumbai and expanded to Delhi, is part of a ₹33,000 crore (≈USD 4 billion) investment in India's digital development. Alongside infrastructure, Google is focused on local partnerships, sustainability research, and its Grow with Google initiatives. Additionally, Google plans to establish a 1GW AI Data Center in Visakhapatnam, powered by renewable energy. This facility will strengthen Google's AI infrastructure in the region, create job opportunities, and promote technological advancements, reinforcing the role of AI in various sectors

These investments reveal that India is no longer just a market for digital consumption but a key production hub for digital infrastructure. This transformation is bolstered by significant commitments from AWS and Google, which are expanding their cloud services in India, reflecting the nation's growing importance in the global digital economy.

The introduction of best practices is a crucial yet often overlooked aspect of foreign direct investment (FDI) in data centers. When firms like AWS or Google open facilities, they implement advanced technologies like power optimization, renewable energy integration, and AI-driven maintenance. They also set high standards that local businesses, whether competitors or allies, must meet, raising industry benchmarks and fostering long-term competitiveness through this transfer of knowledge.

Infrastructure Funds and REITs

Over the past five years, global infrastructure funds have begun treating data centers as core utilities, much like power networks or toll roads, because the digital economy depends on steady compute capacity. Major players such as Brookfield, Stonepeak, KKR, and GIC are building large Indian portfolios. REITs, once focused on offices, now see data centers as long-lease, low-vacancy assets with stable, annuity-like returns. Partnerships like EdgeConneX and AdaniConneX show how equity, long-term debt, and infrastructure bonds now fuel digital infrastructure.

Sovereign Wealth and Pension Funds

Sovereign wealth funds and pension funds, the conservative funds, have increasingly joined the data center universe. Investors such as Abu Dhabi Investment Authority (ADIA), Temasek Holdings, and Ontario Teachers' Pension Plan have invested in Indian and Asian data center platforms.

Their investment is a sign of a greater intelligence: data is the new infrastructure, and its foundations will be as ubiquitous as electricity lines or water pipes used to be. These investments provide patient capital, allowing projects that could take five to seven years to achieve maximum occupancy.

Private Equity and Venture Capital

On a smaller scale, venture capital (VC) and private equity (PE) are funding edge-data center startups, energy-efficiency platforms, and cloud-service intermediaries. While these individually may not hold billion-dollar valuations, collectively they form an ecosystem of innovation in the data center value chain, developing intelligent cooling systems, energy-management software, and localized micro-data nodes for Tier-2 and Tier-3 cities.

This tiered inflow of capital, from hyperscalers into sovereign funds and onto startups, reflects the multi-tier nature of investment that data centers attract and disperse financial participation across geographies and scales.

Real Estate and the Rise of Digital Corridors

Data centers have quietly redrawn the map of India's real estate economy. What were once industrial or warehousing plots are now being transformed into the digital corridors of the twenty-first century. From Mumbai-Thane to Chennai-Sriperumbudur and Hyderabad-Kondapur, land once earmarked for factories has found a new life as the foundation of the digital economy.

From Industrial Plots to Digital Corridors

Municipal planners have started recognizing data centers not merely as commercial properties but as essential infrastructure. Cities like Mumbai, Chennai, and Hyderabad have updated zoning laws to classify them as utilities, opening the door to concessional tariffs, faster approvals, and dedicated "data center parks." This shift has triggered a construction boom in designated zones, where developers and infrastructure companies are racing to meet demand from hyperscalers and colocation providers.

The high-specification requirements of data centers, reliable power supply, proximity to fiber routes, seismic stability, and low flood risk mean that only a handful of urban and peri-urban plots qualify. As a result, scarcity drives up land values. In Mumbai's Airoli, Chandivali, and Navi Mumbai belts, industrial land prices have risen by 30–50 percent in the past five years. Chennai's Ambattur and Sriperumbudur corridors have seen similar surges, with valuations nearly doubling as developers repurpose industrial parks into "data center-ready" campuses.

Redefining Urban Growth

Once an anchor facility begins operations, the ripple effects spread rapidly. Within a few kilometers, logistics parks, substations, and housing complexes start to appear, reconfiguring local landscapes. What happened around automobile plants in the 1990s is now happening again, but with data as the new engine of growth.

In Mumbai's Chandivali and Navi Mumbai zones, old warehousing tracts are evolving into mixed-use developments centerd on data center tenants. Similar transformations are visible in Hyderabad's Kondapur and Chennai's Siruseri belts, where clusters of facilities have turned once-industrial suburbs into thriving hybrid-use technology hubs. These areas are no longer just satellite towns; they are becoming self-sustaining digital ecosystems with their own residential, retail, and logistics economies.

The New Asset Class

This surge has also redefined investment logic in Indian real estate. Developers who once focused on IT parks and SEZs now see data centers as a distinct and highly specialized asset class, capital-intensive, technically demanding, and offering long-term, stable returns.

Major players such as Hiranandani Group (Yotta Infrastructure), Larsen & Toubro (L&T Data Center), DLF, and AdaniConneX have established dedicated divisions for this segment. Their projects reflect the coming of age of India’s digital infrastructure market, where domestic developers compete alongside international hyperscalers to build facilities that meet global benchmarks.

Innovation in Design and Finance

The rise of data centers has pushed Indian real estate into new frontiers of construction. Built-to-suit leasing, where developers construct facilities tailored to tenant specifications, is now common. Joint ownership models allow risk and return to be shared between developers and operators.

Architectural and engineering practices have evolved, too. Design firms have adopted Tier III and IV certification standards, integrated seismic resilience, and embraced sustainability parameters like Power Usage Effectiveness (PUE). This technical upskilling has elevated the entire sector’s sophistication, making Indian developers competitive in the global digital infrastructure market.

Aligning with ‘Digital India’

India's data center growth strongly maps to the country's national initiatives like Digital India, Make in India, and

National Infrastructure Pipeline (NIP). Data centers are the enabling layer for these initiatives to work well.

For instance:

1. Aadhaar, UPI, and DigiLocker rely on massive-scale secure computing power for their success.
2. Manufacturing initiatives like Smart Factories or Industrial IoT depend on cloud analytics.
3. Start-up ecosystems, a highlight of the innovation agenda of the government, rely on scalable cloud platforms.

By developing domestic data center capability, India ensures that its digital aspirations are rooted in local infrastructure and not reliant on foreign computing availability.

Establishing the Backbone for Start-ups and SMEs

Starting a tech company once required heavy spending on servers, networking gear, and licenses. Cloud-based data center ecosystems have flipped that. A small team can now run enterprise-grade applications for a fraction of the old cost. This shift has unlocked enormous economic value because money that once went into hardware can now fuel product development, hiring, and market expansion. By mid-2025, India had over 180,000 active start-ups, and most rely on cloud infrastructure. Fintech, agritech, and many other sectors now use computing power that was once limited to global giants.

Elastic capacity lets young companies scale up during peak demand and scale down to save cash. Lower domestic power costs and cloud credit programs from AWS, Google, and Microsoft strengthen this advantage. SMEs, contributing nearly a third of India's GDP, also benefit. Cloud tools help them digitize operations and compete more effectively. In this way, data centers narrow the gap between large corporations and small innovators.

The Chain Reaction of Confidence

Economists have talked about animal spirits, that intangible confidence that leads entrepreneurs and investors to venture into risk. Data centers, by delivering stability and scalability, amplify those spirits.

When digital infrastructure is solid, entrepreneurs innovate more, investors invest more, and governments scheme larger. The fact that there exists strong data center capacity sends a psychological signal that a country is prepared for the digital age.

This is a trend repeated around the globe. Ireland's tech boom came after hyperscale hubs congregated on Dublin's doorstep; Singapore's transition to a digital economy ran parallel to its emergence as a data center capital. India is now following the same path.

Growth Curve Challenges

The economic potential is huge, but growth is not smooth.

1. Power and sustainability: Increasing energy needs put pressure on utilities and environmental goals.
2. Land supply and price: Limited plots only satisfy technical specifications for hyperscale campuses.
3. Skill gaps: Demand exceeds supply for certified data center engineers, electrical engineers, and operations experts.
4. Regulatory consistency: Investors continue to ask for single national standards for approvals, taxation, and sustainability metrics.

However, these are transitory, the symptoms of an accelerating maturing market rather than structural issues. Each infrastructure revolution goes through an imbalance phase before stabilization. By 2030, global data centers are projected to have a capacity ranging from 134 to 298 GW, while India's capacity is expected to increase to between 3 and 5 GW.

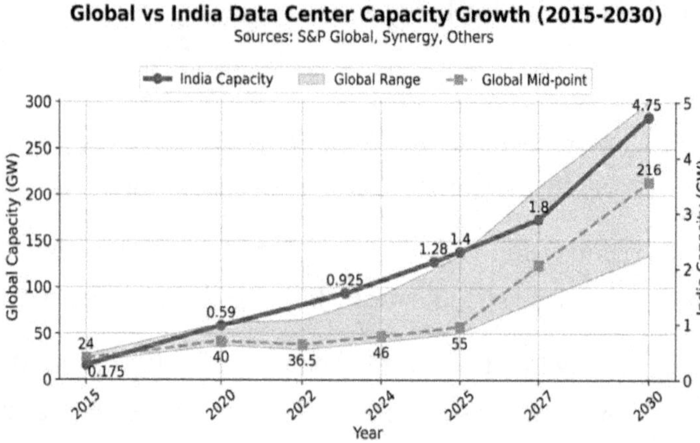

Figure 2.2 Data Center Capacity Over the Last Decade until 2030

The Larger Picture: Infrastructure as Strategy

Countries that invest in data center infrastructure early on reap not only economic dividends but geopolitical influence. Data is both literally and metaphorically powerful, and having the capacity to host, process, and protect it at home determines a country's place in the world order.

For India, this is not just an industry; it's an economic agenda. Every data center campus constructed on Indian ground makes certain that digital value created by its citizens and businesses stays within national bounds, adding to GDP, employment, and innovation.

Employment & Enterprise: Jobs, Skills, and the Emergence of a Digital Workforce

The backdrop of the steel-clad structure of every data center is a human ecosystem. The living web of these enormous digital factories consists of construction crews, electricians, engineers, security guards, software architects, caterers, and even truck drivers.

Once a data center enters a city, it does not import only equipment, but also lives. It alters the way people work, what they study, and what their neighbourhoods resemble. Just like factories revolutionized rural economies in the industrial era, data centers are now revolutionizing urban and semi-urban economies in the digital era.

The Employment Chain: From Groundbreaking to Operations

Data centers create jobs in three linked phases: construction, operations, and ongoing support. The construction stage is the first surge of activity. A typical 30–50 MW facility can employ 500–1,000 workers at its peak, while large hyperscale campuses may involve up to 2,000 people across multiple contractors. Local industries benefit too, supplying materials like cement, steel, ducts, generators, fiber cables, and prefabricated units. Trucking firms transporting heavy equipment add another layer of employment. This 18–30 month phase often supports thousands of households and is now treated by many governments as a priority job-creating sector.

Once operational, the workforce shifts to long-term roles. Medium facilities employ around 100–150 people, and hyperscale sites can support 300–500 specialists across operations, engineering, maintenance, cooling, cybersecurity, and IT services. Indirect jobs in security, housekeeping, catering, transport, and waste management further expand the local economy.

Regular technology upgrades every few years create ongoing employment, making data center jobs more sustainable than most infrastructure projects.

Skill Development and the New Workforce

While the initial industrial revolution gave rise to mechanical trades, the digital infrastructure revolution is giving rise to a hybrid workforce, technical, operational, and managerial.

The Rise of the Data Center Professional

There is a new set of employees: Data Center Professionals, who are trained in specialized fields such as:

1. Electrical and Power Systems Management
2. Precision Cooling and Environmental Monitoring
3. IT Infrastructure Operations (Networking, Storage, Compute)

4. Health, Safety, and Risk Compliance
5. Facility Automation and Digital Twin Technologies

These positions need cross-functional skill sets, combining civil, electrical, mechanical, and IT skills. With India's data center presence spreading wider, the need for such talent has overwhelmed supply.

Training Institutes and Academic Linkages

Realizing the shortfall, various training programs have appeared in various parts of India:

1. National Skill Development Corporation (NSDC) has collaborated with technology companies to develop Data Center Operations Certification programs.
2. Some engineering colleges in Telangana, Tamil Nadu, and Maharashtra are incorporating electives on Data Center Infrastructure Management (DCIM).
3. Private universities such as Schneider Electric University, NTT Global Academy, and Yotta Skill Center provide short-term programs in the skills of technicians and operators.

Such moves are producing a new breed of blue-collar technologists, employees who are comfortable with machines and data and can work between the physical infrastructure and the digital world.

The Indirect Job Ecosystem

Facility Management and Services

Around each giant data center emerges a network of third-party service providers. The firms that specialize in HVAC work, power backup systems, and security services tend to rely on extended contracts with these facilities.

For instance, one hyperscale campus may engage:

- Two full-time cleaning contractors (≈60 staff total)
- A specialized catering vendor (≈30–40 kitchen and delivery staff)
- A multi-year security firm (≈100 guards and supervisors)

Each such provider tends to have multiple facilities, building stable job chains that transcend the walls of a single building.

Transportation and Logistics

With equipment of high value regularly coming and going, secure fleet operations and logistics are essential. Specialist trucking firms now offer vibration-controlled trailers for server transit. The companies have grown explosively along data center corridors, hiring drivers, handlers, and warehouse staff.

Local taxi and shuttle operators also benefit, providing round-the-clock commutes for employees on rotating shifts.

Community and Retail Spillovers

Every cluster of data centers becomes a magnet for local commerce. Cafés, gyms, repair shops, and small restaurants spring up around them, serving employees and contractors. Residential landlords near data center zones report steady rentals, as professionals seek proximity to workplaces.

This chain of second-order businesses creates induced employment, jobs that are not there because of the facility, but because of the individuals that it brings.

Economic Inclusion Through Start-ups

Though direct and indirect employment is a major benefit, the entrepreneurial effect is just as revolutionary.

The Cloud as an Equalizer

Data centers are the infrastructure of the start-up economy. Scalable compute, cloud storage, and stable APIs would have made the explosion of digital entrepreneurship in India impossible.

Almost all fintech payments run on Razorpay, most rides are paired on Ola, major deliveries are followed on Zepto, and depend on data center infrastructure hosted by the cloud. Hyperscale data centers have essentially replaced the physical equipment that characterized industrial capitalism.

For start-ups, this infrastructure removes the barriers to entry. They can:

- Bring products to the world in a matter of days.
- Scale users without hardware concerns.
- Pay as they go (OPEX rather than CAPEX).

This computing utility model, enabled by hyperscale and colocation centers, has democratized innovation.

Regional Start-up Hubs

The availability of large data center clusters has determined where start-ups crop up.

- Mumbai, with its robust cloud network, nurtures fintech and media start-ups.
- Hyderabad, with low latency and affordable computation, is home to start-ups for enterprise-tech and AI.
- Chennai draws SaaS ventures because of its proximity to undersea cables.
- Delhi-NCR brings together cloud infrastructure coupled with venture capital access.

This geographical pattern is not coincidental; it is a reflection of India's data center geography.

Cloud Credit Ecosystems

Initiatives like AWS Activate, Google for Startups Cloud, and Microsoft Founders Hub offer Indian start-ups millions of dollars of cloud credits each year. Translating into computing hours, storage space, and AI toolkits running on local data center servers, such credits help in accelerating experimentation, decreasing the risk of failure, and fostering innovation. Start-up density resulting from this increases the overall economic value of data center infrastructure.

Empowering Small and Medium Enterprises (SMEs)

India's SMEs are undergoing a significant digital transformation. By adopting cloud-based IT infrastructure, these businesses are being integrated into the digital landscape. The emergence of domestic data centers has alleviated challenges like high costs, delays, and security risks. Local cloud hosting has been shown to reduce costs by 20-40 percent, while data localization regulations ensure that critical business information remains within national jurisdiction. This combination of affordability and security has allowed countless SMEs to transition to digital operations smoothly.

Moreover, a thriving ecosystem of SME vendors now supports the data center industry, including cable fabricators, HVAC equipment manufacturers, fire-suppression system providers, and modular construction fabricators. These companies not only enjoy stable demand but also opportunities for upskilling. As a result, the data center economy allows SMEs to engage with technology as consumers and also as contributors in its development and implementation.

Workforce Localization: Beyond the Metros

Prospects for Tier-2 Cities

With data center operators venturing out to Tier-2 towns like Pune, Ahmedabad, Nagpur, Kochi, and Jaipur, jobs are becoming decentralized. Workers who used to go to metros for specialized work no longer require migration.

These initiatives support state-level skilling initiatives, usually in coordination with the states. For instance, Telangana Academy for Skill and Knowledge (TASK) conducts short-duration courses on electricity and network fundamentals for young people in and around Hyderabad's data center parks. Such schemes are now coming up in Tamil Nadu and Gujarat.

Community Stability

The 24/7 operation of data center operations means workers end up settling around facilities, resulting in robust local economies. Data centers offer consistent employment, whereas IT parks can go up or down according to demand from anywhere in the world; servers must be watched over irrespective of business cycles. This consistency lessens outmigration and promotes regional economic resilience.

Shift in Ethical Principles for Work

The Need for Working with Discipline and Precision

Work ethics remain a top concern when working in data centers. Each action, from tightening a cable to keeping an eye on humidity, impacts millions of online users indirectly. The staff internalizes a culture of precision, punctuality, and process discipline.

In that way, data center work is building a new industrial culture of the digital era, based on trustworthiness rather than imagination, on thinking in terms of systems rather than on adapting. These are soft but strong cultural characteristics that will ultimately migrate into other areas of the economy.

The Pride of Invisible Work

Most data center workers refer to their work as "invisible but impactful." While they might not construct goods or greet consumers, their efforts keep the online lives of millions running. This knowledge of one's purpose and unspoken pride is forging a contemporary working-class identity, the guardians of the cloud.

The job and business spillovers of data centers extend beyond the number of jobs in the short run. They are a structural shift of the labour force, from manual to digital, from centralized to distributed, from consumption to production.

Gender and Inclusivity

Many data center operators are designing facilities to be gender-neutral, making them more appealing for women professionals compared to traditional sectors like manufacturing or logistics. Initiatives such as NASSCOM and AWS's "Women in Data Center" are providing technical training and opening new career pathways for women, while efforts to include differently abled individuals in monitoring and documentation roles are gaining traction.

Data centers are generally more inclusive than much of the tech industry due to their need for 24/7 staffing and shift rotations. This environment creates job opportunities for women, differently abled individuals, and local youth. For instance, in Hyderabad, leading data center companies have achieved 30–35 percent female participation in non-IT roles such as control-room monitoring, facility management, and security operations, often prioritizing local hiring over talent from metropolitan areas.

India's Data Center Landscape: Regions, Players, and Tier-2 Horizons

India's transformation into a digital giant has been one of intense smartphone uptake, surging internet penetration, and a constantly growing middle class hungry for digital services. But underlying apps, platforms, and cloud technologies that characterize the new India is something more fundamental: infrastructure.

Although the word "infrastructure" used to evoke images of roads, railways, or power lines, it now more and more refers to something silent: data center clusters, information-processing zones in which information, rather than material, is processed and exported. These clusters are the virtual arteries through which India's new economy circulates.

Mapping India's Digital Geography

India's data center geography is no longer a theoretical one. It is overlaid on cities, corridors, and regions, which are chosen to have an amalgamation of connectivity, power supply, policy support, and business ecosystem.

The national grid is controlled by four big urban agglomerations: Mumbai, Hyderabad, Chennai, and Delhi-NCR. They jointly possess nearly 80 percent of the country's capacity and potential. Surrounding them are a set of Tier-2 cities emerging, Pune, Ahmedabad, Kochi, Nagpur, and many more, each competing to be the next data node in India's digital landscape.

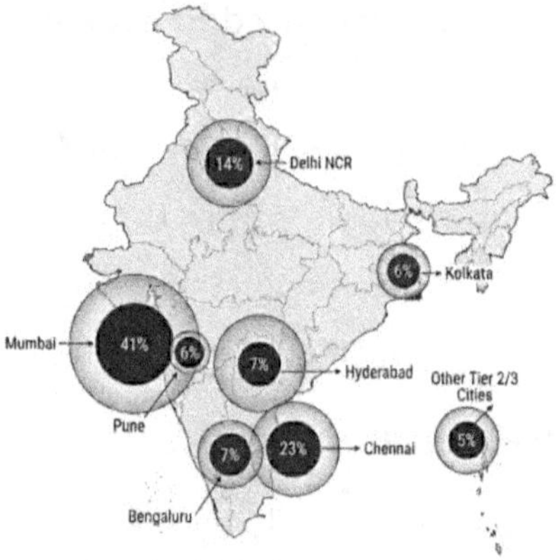

Data Key:
● Solid Dark Circle = Capacity Mid 2025 (% share)
○ Lighter Outer Ring = Projected Capacity by 2030

City	Capacity Mid 2025	Projected Capacity by 2030
Mumbai	~530 MW (41% share)	~1,500 MW
Chennai	~300 MW (23% share)	~900 MW
Delhi NCR	~180 MW (14% share)	~600 MW
Hyderabad	~100 MW (7% share)	~1,100 MW
Bengaluru	~100 MW (7% share)	~300-400 MW
Pune	~90 MW (6% share)	~200-300 MW
Kolkata	~80 MW (6% share)	~150-200 MW
Other Tier 2/3 Cities	~82 MW	300-400 MW

Figure 2.3 Data Center Capacity Over India Map

Mumbai: India's Coastal Gateway of the Cloud

Mumbai has been India's data center hub for more than twenty years, thanks to its proximity to major submarine cable landing stations linking India to Europe, the Middle East, and Southeast Asia. This made it the natural entry point for hyperscalers, shaping the city's shift from commerce to computation. By 2025, Mumbai is set to cross 300 MW of live capacity, with another 400 MW underway across areas like Airoli, Chandivali, Mahape, and Navi Mumbai. Major operators include Yotta, AdaniConneX, NTT, CtrlS, Web Werks, and ST Telemedia. Rising land values and new infrastructure have turned Navi Mumbai into a growing tech corridor.

Hyderabad: The Rising Powerhouse

Hyderabad has become India's fastest-growing data center hub, driven by smart policy and strong infrastructure. Telangana treated data centers as a priority sector early, with its ICT Policy and Data Centre Policy offering single-window clearances, subsidized land, and dedicated high-voltage power. This drew major global players. Adani and EdgeConneX are building a 100 MW campus in Kondapur, while CtrlS is developing a 500 MW mega site in Gachibowli. AWS is investing over USD 4.4 billion, attracted by Hyderabad's central location and low-latency advantage. Engineering talent, vendor ecosystems, and a vibrant SaaS and AI start-up scene reinforce this momentum, earning the city its "Silicon Spine of India" reputation.

Chennai: The Southern Digital Port

Chennai's data center rise is rooted in its strong submarine cable connectivity. Multiple cables land near Tidel Park and Mahabalipuram, linking the city to Singapore, Malaysia, and wider Asia, which makes it a natural gateway for Southeast Asian traffic and a backup route to Mumbai. Tamil Nadu's 2021 Data Centre Policy has strengthened this position through power subsidies, stamp duty relief, and fast-track approvals. The Siruseri–Ambattur–Sriperumbudur corridor now hosts STT GDC India, Sify, Yotta, and AdaniConneX, with over 120 MW live and more than 200 MW planned. Chennai's manufacturing and logistics base, powered by IoT and AI adoption, further accelerates demand for local computation.

Delhi-NCR: The Administrative and Enterprise Core

Delhi-NCR's data center growth is centered in Noida, Greater Noida, and Gurugram. Its strength lies in being close to India's largest enterprise base, key government bodies, and financial regulators. Rising demand comes from banking, insurance, and e-governance workloads that must stay within India under data-localization rules. Uttar Pradesh's 2021 Data Centre Policy accelerated investment through major land subsidies and full stamp-duty waivers. NTT, AdaniConneX, Yotta, Web Werks–Iron Mountain, and Sify are developing large campuses, supported by dedicated power and dark fiber. This momentum is reshaping real estate, turning the Noida–Yamuna Expressway belt into a digital-industrial corridor.

Beyond the Big Four: Emerging Tier-2 Horizons

The economic geography of India is starting to shift. The four metros are at saturation, or near saturation, and are facing pressures of costs, meaning that Tier 2 cities represent the next step for data centers to grow.

Why Tier-2 Cities Are Appealing

There are a number of factors behind this:

1. Lower land and power prices – as much as 40% lower than metros.

2. Supply of skilled but affordable talent.
3. State-level government incentives in Gujarat, Maharashtra, and Kerala.
4. Growing need for edge and regional computing to help 5G and smart-city applications.

The Pioneers

Pune – With its IT heritage and closeness to Mumbai, Pune is developing as a backup hub for disaster recovery (DR) facilities and edge data centers. Its climate is also moderate, which provides energy efficiency benefits.

Ahmedabad – Enjoying Gujarat's pro-investor policies and robust power grid, Ahmedabad is drawing colocation players looking for a central-west India presence.

Kochi – As a cable landing station and maritime hub, Kochi offers unique potential for southern coastal data exchange. The Kerala government's plans for digital parks and renewable energy zones further enhance its appeal.

Nagpur – Located at the center of India geographically, Nagpur is becoming a strategic interconnect hub for latency reduction. The city's future Multi-Modal Logistics Park (MMLP) and digital corridor make it a future-proof location for edge and modular deployments.

The Edge and Tier-2 Synergy

The growth of Tier-2 data centers aligns with the edge computing trend, putting smaller units closer to end-users to keep latency down for use cases such as video streaming, gaming, and IoT. Telecom companies like Jio, Airtel, and Vodafone Idea are already rolling out edge nodes close to 5G towers, commonly in Tier-2 cities where density meets affordability.

This model decentralizes computing, lowers pressure on hyperscale hubs, and brings economic possibilities to secondary cities, establishing a distributed digital economy, not one focused on metros.

Tier 2 and Tier 3 cities are increasingly contributing to India's data center capacity, which is expected to reach between 300 and 400 megawatts by 2030. While their current share is small, it is on a growth trajectory.

The Corporate Frontline: India's Data Center Giants

India's expansion of digital infrastructure is not only the play of international hyperscalers. Domestic giants have joined the game, joining infrastructure development with national ambition.

Reliance Industries (Jio Data Center)

Reliance's digital division, Jio Platforms, has positioned itself as a data-to-device ecosystem player by strategy. Through Jio Data Center and collaborations with Microsoft

Azure, the group is coming up with several facilities integrated with its telecom and enterprise offerings.

These data centers will be 5G edge computing anchors providing ultra-low-latency applications for IoT, autonomous vehicles, and digital entertainment. Reliance's scale, power generation assets, and equity muscle give it a robust position in India's infrastructure strategy.

Adani Group and AdaniConneX

The Adani Group, which has been traditionally present in energy and ports, made an entry in digital infrastructure through its AdaniConneX joint venture with the US-based EdgeConneX. Its plans cover 1 GW of capacity in various Indian cities, i.e., Mumbai, Chennai, Hyderabad, Pune, and Noida, by 2030.

Adani's advantage lies in its integrated capabilities: renewable power generation, logistics networks, and land availability across states. Its approach blends sustainability with scale, aiming for carbon-neutral campuses powered by green energy.

ST Telemedia Global Data Center (STT GDC India)

A ST Telemedia subsidiary, STT GDC India, runs more than 20 centers in 10 cities, one of India's largest colocation players. Its reliability and compliance focus has established it as a go-to partner for BFSI and government customers.

Its hybrid strategy, balancing hyperscale leases and enterprise hosting, is the embodiment of the diversification of India's data center economy, providing capacity for both global behemoths and local enterprises.

Yotta Infrastructure

A part of the Hiranandani Group, Yotta is the intersection of real estate and digital infrastructure. Its celebrated NM1 facility in Navi Mumbai is one of the largest functional Tier IV data centers in Asia, and the firm is going to expand in Chennai, Delhi-NCR, and Kolkata.

Yotta's model is reflective of domestic entrepreneurship in a sector usually replete with foreign capital. It has established "data center parks", integrated ecosystems with cloud, connectivity, and co-working spaces, setting India up to be a digital manufacturing hub within itself.

CtrlS Datacenters

Established in Hyderabad, CtrlS is one of India's native hyperscale vendors. With data centers in Hyderabad, Mumbai, Chennai, and Noida, it has established trust around Tier IV certification and uptime guaranties. CtrlS has outlined a 2 million sq. ft. roadmap of new capacity and carbon-neutral operations by 2030.

The Infrastructure Web: Power and Fiber

The prosperity of data center clusters rests on unseen but essential layers of support.

Power and Energy Infrastructure

Each megawatt of data center capacity requires a continuous power supply. Hyperscale parks have their own substations and dual-feed grids built by the state utility in Maharashtra, Telangana, and Tamil Nadu. Cost-sustainable operators are now entering into long-term Power Purchase Agreements (PPAs) for renewable power.

The Connectivity and Submarine Cables

Mumbai and Chennai are connected to international internet corridors through international submarine cables, such as SEA-ME-WE 5, IAX, and IEX. Onshore in Hyderabad and Delhi, there is fiber redundancy, ensuring that the network does not crash, even if something happens to overseas cables.

Global Parallels & Case Studies: Hyderabad, Northern Virginia, and Singapore

Data centers are not just technical facilities; they are an indication of the way that societies structure capital, talent, and vision in the digital world. If an area becomes a hub of data centers, it is more than an indication of the existence of servers; it is an indication of a nation's readiness for the future.

Around the globe, there are only a few clusters that serve as models of this digital revolution. In India, Hyderabad is remaking its industrial fate with data infrastructure. In the United States, Northern Virginia is now the world's biggest agglomeration of data capacity. And in Asia, Singapore continues to draw global capital as a model of efficiency and policy acumen.

Together, these three stories demonstrate how data center form not only affects economies but also the city and geopolitical identity of countries.

Case Study: Hyderabad, India's Emergent Data Center Cluster

Hyderabad has reinvented itself. Once known for biryani, pearls, and software services, the city now sits at the heart of India’s digital infrastructure boom. Its journey from an outsourcing hub to a data-infrastructure powerhouse is the result of clear vision, strong policy, coordinated industry effort, and focused urban planning. The turning point came in the late 2010s, when Telangana recognized that data was becoming the new industrial raw material. It didn’t need ports or freight networks. It needed reliable power, strong connectivity, and available land. This insight shaped the ICT Policy of 2016 and the Data Centre Policy of 2021, positioning Hyderabad as an ideal site for hyperscale campuses.

The state’s policy framework offered speed and predictability. Investors received single-window approvals, land in key IT corridors, and access to dedicated substations. Developers were allowed to source renewable energy directly, which kept operating costs low and reduced environmental impact. This fast conversion of policy into physical projects became Hyderabad's biggest competitive edge.

That momentum anchored a large cluster stretching across Gachibowli, Kondapur, and Nanakramguda. Massive developments by CtrlS, AdaniConneX, Yotta, and AWS reshaped the skyline. AWS committed USD 4.4 billion through 2030. CtrlS built a two-million-square-foot Tier IV campus. AdaniConneX is rolling out a modular 100 MW facility powered by clean energy. Their presence drew contractors, vendors, and training institutes, creating a self-sustaining ecosystem.

The surge has generated more than 30,000 jobs and lifted commercial real estate and hospitality demand. It has also reset the city's identity—from a software exporter to a center of infrastructure innovation. With rising renewable adoption and new green-energy incentives, Hyderabad is shaping a model of how regional governance can steer national-scale digital growth.

Case Study: Northern Virginia, The World's data center Capital

About 50 kilometers from Washington, D.C., sits Loudoun County in Northern Virginia, known worldwide as "Data Center Alley." More than 275 data centers operate here, handling roughly 70% of global internet traffic during peak hours. Its rise began in the 1990s, when parts of the ARPANET network were retired, and telecom carriers set up exchange points outside the capital. The region's strategic location, political stability, and reliable power laid the foundation for extraordinary growth.

What truly transformed Northern Virginia was long-term planning. Local authorities carved out zoning for data centers, built high-capacity substations, and encouraged multiple carriers to lay redundant fiber. State incentives, including sales-tax breaks on equipment, attracted global operators. This created a reinforcing cycle: each new facility drew more networks, which in turn attracted major cloud companies seeking low latency and deep connectivity. Ashburn eventually became to data what Wall Street is to finance.

The economic impact is enormous. Data centers contribute over USD 7.5 billion to Virginia's GDP each year, support more than 45,000 jobs, and generate upwards of USD 600 million in property taxes, funding schools, transport, and public services while keeping personal tax rates low.

Such a scale brings scrutiny. Northern Virginia consumes 3–4 GW of power, prompting sustainability concerns. Operators have responded with renewable energy agreements, liquid cooling, and heat-reuse ideas. Companies like Microsoft, Amazon, and Meta have pledged carbon neutrality by 2030.

The key lessons for India are clear: build infrastructure early, ensure policy stability, encourage clustering, and keep communities involved.

Case Study: Singapore, The Asian Magnet for Digital FDI

Singapore's rise from port city to global data hub is the result of smart geography matched with disciplined policy. Sitting at the intersection of major submarine cable routes, it connects Southeast Asia, Europe, and the Pacific. But its real edge comes from how it manages technology, investment, and regulation. Since the 1990s, the country has treated trust, reliability, and speed as pillars of its digital economy, weaving them into both infrastructure and governance.

The Infocomm Media Development Authority (IMDA) and Economic Development Board (EDB) created a predictable environment with single-window approvals, transparent environmental norms, and consistent power-allocation rules. This attracted global giants such as Google, Amazon, Meta, Microsoft, and Equinix, pushing Singapore toward an expected 1,000 MW of operational capacity by 2025, remarkable for its size.

Rapid growth also brought challenges. Energy use and land constraints triggered a temporary pause on new data center construction in 2019. Rather than slowing progress, this led to a reset. IMDA and EDB introduced a Green Data Center Roadmap targeting higher efficiency, renewable energy, and vertical, space-saving designs. New facilities are now more compact and sustainable.

The sector contributes over SG$4 billion annually and supports thousands of jobs, but its influence extends further. By hosting digital infrastructure for global firms, Singapore has become the digital nerve center of ASEAN. Its model shows how clear policy, sustainability, and strategic geography can create both economic strength and regional leadership.

The Broader Narrative: Data Centers as Global Economic Engines

Across continents, the data center economy is transforming the way nations conceive of development. Ports, railways, and oil refineries characterized geopolitical power in previous centuries. Today, the dominance over data flows is what establishes economic sovereignty.

Countries with large data hubs are magnets for foreign capital, skilled employment, and digital ecosystems; those without risk becoming consumers, not producers, of digital value.

In this regard, data centers are not mere private investments, but public assets in disguise that bring nations resilience, self-sufficiency, and leverage in international digital commerce.

India in the Global Context

India occupies a singular combination of size and aspiration. With 1.4 billion digital citizens, a burgeoning start-up ecosystem, and a digitally focused government agenda, the need for local data processing is vast. The decade ahead will witness:

1. Around 2 GW of new capacity has been added across the country.
2. $30–35 billion of overall investments in hyperscale and colocation.
3. Appearance of regional clusters coupled with renewable energy and edge networks.

India can not only construct data center capacity but also digital sovereignty by following global leaders, yet forging its own way, to ensure that value is created in India out of the data generated by Indians.

The Economic Multiplier Revisited

Synthesizing the international evidence, the multiplier effect of the data center is even more evident:

1. For each $1 billion invested, a total economic output of $1.8–$2.2 billion is estimated.
2. For every 100 direct jobs, 400–500 indirect and induced jobs are sustained.
3. For each megawatt of capacity, close to ₹1,000 crore in lifecycle ecosystem expenditure takes place.

These figures explain why countries now fight as hard for data center investments as they used to for factory plants.

The Architecture of Prosperity

The data center story is, in essence, the story of modern civilization's unseen architecture, the infrastructure on which our economies, our professions, and even our fantasies now depend.

From Hyderabad's new-age corridors to the server farms of Loudoun County, from Singapore's vertical data skyscrapers to Tier-2 ambitions of India, the message is ever the same: where the data flows, growth ensues.

As India builds its digital infrastructure, it is not only constructing data centers, but it is also building the pillars of a new economic order. These soundless structures, whirring along in air-conditioned silence, are the factories of the future. And in their sheen lies the hope of an interconnected, prosperous, and self-sufficient India.

From Growth to Governance

Growth, all its dynamism notwithstanding, sooner or later requires to be structured. The thrum of servers and shine of new campuses might whisper the tale of advance, but underneath that advance comes a greater question: who rules the grid of the information age?

As India's data center economy grows to gigawatt scales, its challenges are moving away from building to coordination, from how much we can build to how well we can manage it. The subsequent chapter, then, shifts from expansion economics to administration ethics. It considers how laws, incentives, and institutions make digital power operate in the service not only of profit, but of public good.

Chapter 3

Policy, Governance and Regulation

"The purpose of regulation is not to control technology but to civilize it." - Edmund S. Muskie.

The Architecture of Control: Why Data Needs to Be Governed

When countries industrialized during the nineteenth century, they constructed laws about factories, workers, and wages. When they globalized during the twentieth century, they constructed rules about trade and finance.

Now, in the twenty-first century, as the world becomes digitized, they are discovering that information too requires its own legal framework, its own ownership, movement, and responsibility rules.

Data has become an asset and ammunition. It powers business, drives economies, enables citizens, but also subjects them to exposure. Each byte that travels through a data center carries not only value, but vulnerability. That's why governance isn't a bureaucratic add-on; it's the digital state's central nervous system.

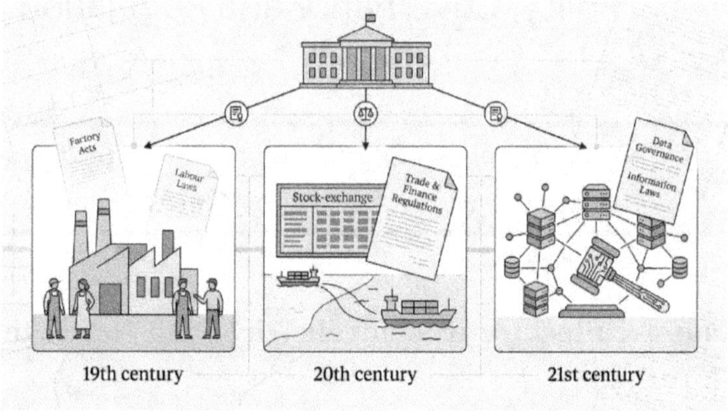

Figure 3.1 Governance Over Decades

The Governance Imperative

For most of the past two decades, the digital economy functioned in a policy vacuum. Innovation outran law, and technology outpaced institutions' ability to keep pace. The playground of experimentation has progressed into a structure of immense significance to national security, economic stability, and personal privacy.

Data centers have become strategic assets in this new world and are not neutral infrastructure. They house financial transactions, health records, government databases, defence communications, and corporate intelligence. The failure or misuse of these systems can ripple through economies and countries.

Governance in the data center ecosystem, then, is not a choice; it is sine qua non (a matter of existence). It dictates:

- Who can store and process data?
- Where that data can be located.
- How it can be accessed, shared, and monetized.
- What responsibilities fall on operators and governments alike?

The conversation about data center governance is ultimately a conversation about sovereignty and safety in the digital age. The more a nation depends on digital infrastructure, the more it must define the principles guiding its use.

Data as a Sovereign Resource

The New Oil Analogy, and Its Limits

The phrase "data is the new oil" captures part of the truth about its economic power. But data is not like oil in that it is regenerative, omnipresent, and intimate. It doesn't only power industries; it also shapes individuality.

This hybrid character of data, economic, private, and human, complicates its proper governance. A liter of oil does not tell you who you are; a terabyte of information about you might do it. So, whereas industrial regulation was constructed on scarcity, data regulation needs to be constructed on ethics, consent, and control.

The Rise of Data Sovereignty

As cloud providers globally grew, nations realized that citizens' data was being stored outside their jurisdiction. This gave rise to the concept of data sovereignty, which is the notion that data created within a nation must be subject to its laws, in spite of being hosted by a foreign entity.

India, as a country, has also now embraced this idea. Such an exercise of sovereignty is reflected in the Draft Personal Data Protection Bill (2019), and, later on, in the Digital Personal Data Protection Act (2023). While its implementation is still pending, it does offer a good ground for data sovereignty, and we can hope for a better future. This Digital Personal Data Protection Act requires that important personal data be processed or stored in India, and this is a decisive step towards making India self-sufficient when it comes to digital data and systems.

Information to Infrastructure

By making data a sovereign asset, India essentially categorized data centers as strategic properties. In the same way as ports or power plants are controlled by the laws of the country, data centers must be obligated to operate within the governance framework that would ensure the safety of the population and the ethical application of the data.

The Government as Enabler and Regulator

Dual Role, Delicate Balance

The data center economy entails a paradox of governance. The state must simultaneously fulfil two responsibilities: the role of an innovation enabler and the role of a protector of public interests. Too much regulation can hinder growth, while too little can lead to chaos.

This balance is achieved through the following:

1. Support frameworks should be established to encourage investors to invest in the country. This includes offering tax incentives, subsidies, and streamlined approval processes.
2. It is also essential to ensure regulatory safety by implementing laws that protect the population, focusing on privacy, cybersecurity, and data localization.
3. Additionally, it is important to build strong institutions that can effectively manage and adapt to a rapidly changing ecosystem.

India is making strides in addressing its challenges, and the focus is on enhancing collaboration among various ministries, state governments, and the private sector to streamline efforts effectively. Digital infrastructure is not limited to a single department; it encompasses various areas such as communications, IT, energy, environment, and finance.

Adaptive Policy Frameworks for Better Governance

Unlike traditional regulations that change slowly, digital governance must be flexible and adaptable. The rapid advancements in technology, such as cloud computing, AI, and blockchain, can make static laws outdated in just a few years. In India, the approach has shifted towards adaptive policy frameworks that allow for modifications through notifications and executive orders, rather than relying solely on lengthy legislative processes.

This model ensures that governance remains aligned with technological progress while maintaining trust and accountability.

Ethical Considerations for Efficient Governance

Every system of governance involves a balance between individual freedom and necessary regulation. This balance is further complicated by the rise of the digital world.

As we navigate this landscape, important questions arise: Should individuals retain ownership of their personal data? Can governments implement measures to safeguard security while managing the flow of data across borders? Is it fair for companies to generate revenue from data analysis, and how can they ensure they do so with clear and informed consent?

These questions highlight the ethical considerations that shape modern governance and the ongoing dialogue about protecting individual rights while promoting innovation and security.

The Freedom of Innovation

Innovation often thrives in uncertain environments when it isn't held back by strict regulations or guidelines. The early IT entrepreneurs created products at a more rapid pace than the regulators could comprehend them, and it was this agility that created international behemoths. However, when violations, fake news, and algorithm favoritism grew, societies began to demand responsibility.

Regulation, once viewed as a limitation, is now increasingly seen as a guiding force that drives innovation rather than stifling it.

The Need for Security and Control Over Data

National security interests necessitate the implementation of control measures. Sensitive information, defence communications, financial networks, and healthcare databases cannot be left open to extraterritorial threats. This has prompted India, like others, to prioritize data localization and national control of digital infrastructure.

The Ethics of Data Stewardship

The real question is not whether data should be controlled, but rather how it should be managed and who should have that authority. Governance must ensure that control does not become surveillance and that protection does not infringe upon freedom.

Digital governance is therefore an ethical responsibility, ensuring that authority over information is transparent, accountable, and ultimately answers to citizens.

The Institutional Architecture of Governance

Ministry and Agency Ecosystem

India's data center governance is dispersed over several institutions:

1. The Ministry of Electronics and Information Technology (MeitY) determines national digital policy and standards.
2. The Department of Telecommunications (DoT) regulates connectivity and interconnection standards.
3. The Ministry of Power governs energy supply to large campuses.
4. The Central Electricity Authority (CEA) controls energy efficiency and redundancy in hyperscale campuses.

5. The State IT Departments address local policy implementation, allocation of land, and fiscal incentives.

This multi-level system demands harmony among economic incentives, legal safeguards, and technical standards.

The Role of NASSCOM, DSCI, and Industry Bodies

Non-state actors such as NASSCOM (National Association of Software and Service Companies), DSCI (Data Security Council of India), Cyber Security Association of India (CSAI), Digital Empowerment Foundation (DEF), The Energy and Resources Institute (TERI), and National Solar Energy Federation of India (NSEFI) serve as intermediaries between the state and the private sector. These organizations prepare voluntary codes of conduct, issue recommendations on cybersecurity frameworks, and shape policy through consultation.

Their work reflects a developing model of co-regulation, not top-down imposition of governance, but together with industry, government, and civil society, co-created by conversation.

Trust as Infrastructure

At its core, governance is about trust, not in the technology, but in the institutions that govern it.

A data center can promise uptime and redundancy, but ultimate resilience lies in public trust that data will be handled fairly, securely, and openly.

Without trust, digital infrastructure is empty, a technological achievement with no social credibility.

With trust, even intricate regulatory frameworks can run smoothly, because citizens regard governance as protection rather than intrusion. This trust is established with the rolling out of the DPDP Act, 2023, which after its implementation will be something that will further change the landscape of data centers in India.

The Turning Point

India's journey toward digital infrastructure has come to a crucial point of inflection. Having established the capacity and drawn investment, the next challenge is to make growth stick to governance.

The data center industry is now where India's financial industry was in the 1990s, with all opportunities but requiring regulation that is up to its pace and size.

India's shifting policy environment, extending from data protection to investment guidelines and cybersecurity legislation, will set not only how the industry expands, but also how responsibly it expands.

India's Policy Evolution and Regulatory Landscape

If the initial phase of India's data center saga was one of building and investment, the second one is one of codification, providing form to expansion through legislation, standards, and collaborative government. Regulation in this context is not a brake on advancement but its steering wheel; in its absence, velocity can degenerate into drift.

The Roots of Digital Policy in India

India's digital policy journey started well before the terms cloud computing or hyperscale featured in the business lexicon. In the 1990s, the Information Technology Act (2000) introduced a skeletal blueprint for electronic transactions, digital signatures, and cybercrimes. It was reactive and a legal safety net for an emerging internet economy.

By the mid-2010s, though, the country had transformed. Citizens were now data generators thanks to the spread of smartphones, and online services were essential infrastructure. The policy conversation evolved from connectivity to control, capacity, and confidence.

That transformation created three interlocking policy curves:

1. Digital Infrastructure – transforming India into a data hosting and transmission hub of the world.

2. Data Protection and Localization – establishing national jurisdiction over digital assets.
3. Cybersecurity and Trust Frameworks – safeguarding data, institutions, and citizens.

The National Digital Communications Policy (2018)

The initial large-scale effort at combining infrastructure and governance arrived in the form of the NDCP 2018, which envisioned telecommunications as the pillar of a digital society. It identified data centers and cloud ecosystems as essential infrastructure warranting policy support at least comparable to highways or airports.

Salient provisions were:

1. Infrastructure Status – enabling greater access to finance and insurance for data center developers.
2. Streamlined Right-of-Way regulations for fibre and power connections.
3. Facilitating Cloud Services as part of India's export basket.
4. Encouraging data center parks through public-private partnerships.

Whereas the language of the NDCP was aspirational, it established the precedent: data was now a public-interest space, not a private good.

The Draft National Data Centre Policy (2020) and Subsequent State Frameworks

Seeing the necessity of a country-specific national roadmap, the Ministry of Electronics and Information Technology (MeitY) made available the Draft Data Centre Policy 2020.

Its goal was trifurcated:

1. Establish an ecosystem of data centers via fiscal and non-fiscal support.
2. Facilitate sustainable and energy-efficient architecture.
3. Develop India as an international data-storage destination under the vision of "Digital India."

The note triggered action even before adoption. Several states, Maharashtra, Tamil Nadu, Telangana, Uttar Pradesh, and Gujarat, came up with their own data center policies, adapting incentives to local conditions.

Maharashtra

Early bird; made data centers essential services to ensure 24/7 power.

Provides 100% exemption from electricity duty and a 50% duty concession.

Navi Mumbai and Pune emerged as a magnet for domestic and international operators.

Tamil Nadu

The policy connects digital infrastructure to industrial corridors.

Offer twin-power-feed guarantee and green-energy rebates.

Fosters academic collaboration for skilled human resources.

Telangana

Most investor-friendly, with single-window clearance and land incentives.

Promotes renewable-powered campuses and local employment.

Uttar Pradesh

Aggressive fiscal package: 50% land discount, 100% exemption of stamp duty, and a capital grant for infrastructure in Noida-Greater Noida.

Arrived at the concept of Data Center Parks, pre-approved clusters with exclusive substations.

Collectively, these frameworks have made regulation competitive federalism, the states competing not only for factories, but for server farms.

Digital Personal Data Protection Act (2023)

The DPDP Act 2023 is possibly India's most impactful regulatory piece of the puzzle, and the nation's first all-encompassing data-protection legislation. It redefines privacy as a statutory right and enshrines how personal data can and must be collected, processed, stored, and shared.

Key features:

1. Consent-based processing – organizations have to get clear consent from users.
2. Purpose limitation – information can be used only for the purpose for which it was gathered.
3. Right to correction and erasure – one can request rectification or erasure of data.
4. Cross-border rules – sensitive and critical data types may have to be stored locally.
5. Formation of the Data Protection Board of India – a quasi-judicial tribunal to adjudicate violations.

To data center operators, the Act means strict compliance. Sites have to guarantee auditability, traceability, and breach-notification procedures. But more than compliance, it instills a philosophical change: data stewardship as a public trust, rather than a commercial service.

The Interplay Between Central and State Governance

India's model of governing digital infrastructure is federal but interdependent. Macro-policy and legal frameworks are established by the center, while states give land, power, and implementation. Interdependence can be a cause of both synergy and friction.

Synergy: Central incentives (such as infrastructure status or export benefits) get leveraged when states offer local tax relief.

Friction: Overlapping jurisdictions can slow down clearances, for instance, environmental clearances may be with central agencies, while conversion of land may be with states.

To facilitate coordination, MeitY has introduced the National Single-Window Digital Platform, enabling investors to monitor approvals across departments, a small but important step towards bureaucratic transparency.

Fiscal Incentives and Taxation Frameworks

Data center projects are heavy on capital and long on the payback period. Governments have thus depended on fiscal baits to attract investment.

Common incentives are:

1. Stamp-duty exemption (25–100%) on land deals.
2. Up to 20 years of tax exemptions.

3. Electricity-duty waiver for 5–10 years.
4. Refund of SGST on infra material.
5. Subsidy for interest in integrating green energy.
6. Power-tariff concessions on renewable-energy procurement.

At the national level, an "Infrastructure Sector" classification of data centers in 2020 opened the door to long-term institutional financing, enabling projects to borrow cheaper funds through bonds and pension schemes.

In 2023, the government went on to state that data center services are eligible for 100% FDI under the automatic route, a landmark move that boosted global investors' confidence.

Governance in Practice: Operational Resilience and Cybersecurity

Policy on paper is only as effective as its implementation on the ground. India's cybersecurity governance machinery has its roots in the Indian Computer Emergency Response Team (CERT-In), which commands breach reporting and threat containment directives.

Some of the key directives are:

- Forced reporting of cyber incidents within six hours.
- Storage of system logs for a minimum of 180 days in India.

- KYC customer verification for cloud and VPN services.
- Conformity with the National Critical Information Infrastructure Protection Center (NCIIPC) of critical sectors, finance, power, telecom, and government networks.

In the case of hyperscale and colocation providers, it implies consolidated Security Operations Centres SOCs), real-time threat analysis, and continuous auditing. The result is a progressive yet cumulative shift in the direction of reactive cybersecurity to resilience by design.

Mechanisms of Accountability and Transparency

The government should not be an opaque organization. The Data Protection Board, when operational, will be in a position to impose fines and take remedial measures. At the same time, the Right to Information Act and Public Accountability Portals expose policy implementations on the state level to the monitoring of people.

Some of them, anticipating international scrutiny, have made voluntary ISO 27001, ISO 27701, and SOC 2 certifications, and made a connection between Indian compliance and international assurance standards.

The Challenge of Regulatory Convergence

Digital infrastructure is rarely accommodated in well-defined policy silos.

A single data center initiative can initiate:

- Environmental clearance (Ministry of Environment).
- Building and fire approvals (Urban Development).
- Energy permissions (State Power Utilities).
- Telecom licensing (DoT).
- Cyber-security audits (MeitY / CERT-In).

This diversity can be a source of friction, especially with foreign investors accustomed to more such dispersal. Seeing this, the government is mulling a National Digital Infrastructure Authority, a single nodal agency to coordinate rules across ministries. If implemented, it might do for data centers what SEBI has done for the capital markets: make compliance consistent.

Institutional Maturity and the Role of the Judiciary

India's judiciary has emerged as an unlikely partner in digital governance. From the Puttaswamy Judgment (2017) upholding privacy as a constitutionally protected right, to decisions on intermediary liability and data interception, courts have progressively set the moral boundary of regulation.

These judicial norms typically advise executive agencies while writing rules. They also warn policymakers against the notion that governance can and cannot be entirely an issue of efficiency, but governance should always be an issue of protecting the values of the constitution, dignity, autonomy, and fairness.

Public-Private Partnership: Emerging Administrative Spirit

The best development in the regulatory environment in India can be seen in the emergence of collaborative governance. Associations of industries, think-tanks, and universities are currently involved in direct policymaking by way of consultations, white papers, and pilot projects.

Examples are:

1. NASSCOM's Data Protection Taskforce, which helped refine the DPDP Act's operational clauses.
2. DSCI's Cybersecurity Framework, developed jointly with MeitY.
3. Cyber Security Association of India (CSAI), which collaborates on cybersecurity policies at the national and state levels to secure digital infrastructure
4. Power and Green-Energy Partnerships with operators and the Bureau of Energy Efficiency to establish sustainability standards.
5. The Energy and Resources Institute (TERI) and the National Solar Energy Federation of India (NSEFI) are

advocating for green energy use and policies for Eco-friendly data center growth.

This collaborative effort is a sign of an advanced state, one that is secure enough to share the table and not dictate from there.

Governance That Goes Beyond Compliance

Good governance isn't checklists. A nation may have exemplary rules and yet fail if its institutions are short on capacity, transparency, or will. India's next challenge is converting policy surplus into execution mastery:

- Automate approval processes to remove bottlenecks.
- Educate bureaucrats on emerging technologies.
- Enhance local regulatory offices with technical personnel.
- Establish independent review boards for major infrastructure audits.

These are unglamorous measures, but they make statutes systems.

The Cultural Layer of Regulation

Regulation also transforms culture. Data center developers today design projects with legality as a design parameter, power redundancy, data localization, renewable supply, and fire protection built into the master plan, not added after the fact. Law has turned into a type of engineering discipline.

At the same time, the citizens are building a culture of data rights consciousness, requesting consent notices, privacy dashboards, and secure transactions. The outcome is a quiet social evolution: governance is not perceived as an obstruction anymore, but as a provider of fairness in the digital economy.

India's policy evolution reveals a clear trajectory: from enabling growth to institutionalizing trust. Underlying are the establishment, infrastructure situation, financial incentives, confidentiality regulations, and cybersecurity standards. Their second breakthrough is to coordinate and unify them into one adjustable regulatory fabric capable of withstanding society's changes in the coming technologies, such as AI, quantum computers, and edge networks.

The government will not only protect what exists when India enters this next chapter, but it will also make what is possible.

Striking a Balance Between Innovation, Sovereignty, and Global Alignment.

Each country trying to tap the digital economy is confronted by the same paradox: how to foster the innovation of creativity without compromising the discipline of regulation.

For India, the paradox is compounded by scale. With over a billion digital users and an investment-hungry economy, neither unbridled freedom nor crippling control can be afforded. What it needs instead is a balanced rule, a system that encourages experimentation with a sense of grounded responsibility.

The Need for Balance

The industrial age brought about trade and tariffs, wages and growth, and production and pollution control being the key issues in countries. In cyberspace, the pendulum has swung toward privacy over progress, sovereignty over openness, and control over innovation.

Such governance today has to respond to three interconnected questions:

- How can India keep its citizens' data secure while attracting cross-border capital?
- How can it promote start-ups and cloud services without undermining national security?
- How can it do so without becoming subservient to them?

These are not policy matters; they are questions of civilization, deciding the course by which India will conceptualize digital power in the coming decades.

Innovation and the Regulatory Mindset

Regulation as Enabler

The most advanced perspective on regulation views it not as a hindrance but as an innovation platform. Where there are clear rules, entrepreneurs and investors feel secure. They understand the limits within which risk may be taken. India's recent history confirms this.

The Digital Personal Data Protection Act assured global clients that privacy could be enforced, and outsourcing and migration to the cloud were encouraged.

The 100% FDI permit under the automatic route made capital flows easy.

State single-window portals shave months of red tape off data center builders.

Every regulatory refinement converted prudence into confidence, turning the act of governance into a driver for innovation.

Sandboxes and Experimentation

Acknowledging that technology changes at a speed that outstrips law, regulators have taken to introducing regulatory sandboxes, safe spaces where new products may be tried out under close watch. The Reserve Bank of India led the way for fintech; MeitY is looking at identical models for cloud computing services, artificial intelligence, and cybersecurity products.

These sandboxes are a cultural milestone: the state learning to collaborate with innovators instead of keeping them at arm's length.

The Sovereignty Dimension

The Logic of Localization

Data localization has been the foundation of India's digital sovereignty. It protects sensitive personal and financial information within national jurisdiction, available to local law enforcement and regulators.

Critics claim that localization raises costs; proponents reply that it builds resilience and national security. The middle ground is "localize critical, distribute non-critical", a tiered strategy with flexibility whilst protecting strategic assets.

Strategic Autonomy in Infrastructure

Sovereignty also lies in who develops and owns the infrastructure. By cultivating home-grown data center firms, Yotta, CtrlS, AdaniConneX, India ensures that not every digital foundation relies on foreign parties. This diversification reflects India's time-tested doctrine of strategic autonomy in defence and energy: engage globally, but rely nationally.

Trust and Global Cooperation

True sovereignty does not concern isolation. Modern data streams are transnational in nature; even storage locality often requires cross-border backup, analytics, or AI training. The strategy of India has thus shifted to trust-based openness and has entered into data-transfer agreements with countries with similar privacy standards, but with the right to restrict flows to countries with different privacy standards.

This is the philosophy that forms the core of the Indian negotiation policy on international digital-trade negotiations.

Global Alignment: Learning Without Imitation

The European Union – GDPR's Moral Framework

The General Data Protection Regulation (GDPR) set the new standard for privacy. Its fundamental principles, lawfulness, transparency, purpose limitation, and right to be forgotten, have shaped legislation on other continents, including India's DPDP Act. But India improvises, not imitates: it accords equal weight to digital inclusion and compliance convenience, aware that over-stringent legislation can weigh down SMEs and start-ups.

The United States – Market-Led Governance

America's model is focused on innovation and auto-regulation. Sectoral legislation (such as HIPAA for healthcare or GLBA for the financial sector) regulates particular areas, whereas competition in the marketplace ensures accountability. India adopts selectively from this model through promoting industry codes of practice under NASSCOM and DSCI, yet has a more robust state involvement to safeguard citizens within a less developed marketplace.

The ASEAN Model – Regional Coordination

Southeast Asia's strategy, exemplified by Singapore, Malaysia, and Indonesia, strikes a balance between global capital and rigorous data-residency regulations. India has a similar regional philosophy: welcoming investment but guarding against reliance. Strategic discussions between the Infocomm Media Development Authority (IMDA) of Singapore and MeitY examine standards interoperability, demonstrating why India is determined to become integrated into the new Asian Digital Commons.

Global Case Studies

Case Study: Europe's Digital Single Market – Turning Fragmentation into Strength

Europe didn't arrive at digital harmony by accident. It spent more than a decade confronting a basic challenge: twenty-seven nations, each with its own rules, standards, privacy regimes, and definitions of what "digital" even meant. In the early years, this patchwork slowed innovation. A company compliant in France could easily fall foul of rules in Germany or Spain. Consumers were living in a divided digital world where data protection, online commerce, and cybersecurity all operated on different national clocks.

The Digital Single Market (DSM) emerged as Europe's effort to stitch these pieces together. Brussels treated data as a continental resource and built a regulatory spine that everyone could plug into. The General Data Protection Regulation (GDPR) did more than protect privacy. It created a shared language of rights and responsibilities. The NIS Directive did the same for cybersecurity, forcing critical infrastructure operators to follow common security standards across borders.

The process wasn't smooth. Member states argued, industries protested, and early drafts went through countless rewrites. But Europe understood a simple truth: digital markets grow only when trust and predictability scale. Harmonizing standards meant that a start-up in Vilnius could serve customers in Madrid without navigating twenty-seven bureaucracies.

India's federal structure is not very different from Europe's political mosaic. States like Maharashtra, Telangana, and Uttar Pradesh already have their own digital policies and incentive regimes. As India's data economy matures, the European experience shows that convergence eventually becomes essential. The DSM is a reminder that aligning rules across jurisdictions isn't about controlling innovation; it's about giving innovation a stable foundation on which to multiply.

Case Study: California's CCPA – A Sub-National Experiment That Shifted a Nation

The United States never adopted a single federal privacy law, so the battleground moved to the states. California stepped in first with the California Consumer Privacy Act (CCPA), a law that surprised many by becoming a national benchmark. The CCPA wasn't born from bureaucracy; it originated as a citizen-led push driven by growing discomfort with how tech companies collected and monetized data. California's lawmakers realized they had to respond before the state's booming tech economy pushed boundaries faster than society could react.

The CCPA gave Americans a vocabulary they didn't previously have. People learned what a "data broker" was, what it meant to "sell" data, and how to request deletion. For companies, the law forced the kind of transparency they had long avoided. Suddenly, privacy policies were rewritten, data flows were audited, and compliance teams grew from token groups to substantial departments.

The national impact was clear within two years. Virginia, Colorado, and a dozen other states built their own laws, often copying California's template. The federal government, reluctant and slow for years, began debating national legislation again simply because state experimentation made the old silence impossible.

For India, there's a direct parallel. States are already crafting their own digital incentives and data frameworks. That experimentation is healthy and even necessary. Sub-national policy often acts like a prototype. Some states take bolder steps, others hold back, and eventually the country gets a clearer sense of what works. The American story shows that federal architecture often grows out of local trials, not the other way around.

Case Study: Australia's My Health Record – A Policy Failure That Became a Lesson

Not every digital policy becomes a showcase success. Australia's My Health Record project is one of the most cited cases of what can go wrong when ambition outpaces communication and public trust. The idea seemed sound: create a centralized digital health record for every citizen and give doctors faster access to medical histories. In theory it would improve care, reduce duplication, and save lives. On paper, it was hard to oppose.

The execution told a different story. The government initially made the system opt-out rather than opt-in. Millions of Australians woke up to find that their health data had been placed in a centralized database without their explicit consent. Privacy advocates raised alarms, cybersecurity experts pointed out vulnerabilities, and citizens grew angry at the sense of being coerced into a system they didn't fully understand.

Security incidents didn't help. Even though breaches were limited, the perception of risk became the reality. Public trust evaporated. The government eventually had to overhaul the policy, change access rules, tighten penalties, and launch a massive communication campaign to win back credibility. The project survived, but only after years of reputational damage.

This case offers a clear lesson for India as it builds digital health frameworks, data exchanges, and public-sector digital infrastructure. People are willing to accept large-scale data systems only when they feel respected and informed. Consent cannot be a legal technicality; it must be a lived experience. Transparency cannot follow outrage; it must precede it. Australia's missteps remind policymakers that trust, once lost, is expensive to rebuild, and that the success of digital public goods depends less on technology and more on how citizens perceive the intent behind it.

The Triad of Future Governance: Security, Sustainability, and Inclusion

Security

Cyber threats are geopolitically-minded. Grid, payment network, or health database attacks can shut down societies. India's efforts toward enforcing mandatory reporting of incidents, threat-sharing practices, and AI-powered cyber-defence demonstrate an awareness that digital safety is as important as border safety.

Sustainability

Energy use is the new regulatory frontier. For every megawatt of data center load, decision-makers have to balance the carbon footprint. The Bureau of Energy Efficiency, in collaboration with operators, is creating Green Data Center Guidelines, benchmarks on renewable sourcing, cooling technology, and heat-recovery systems. Carbon credits might be added to compliance certificates, along with uptime ratings, within the next decade.

Inclusion

Governance also needs to ensure that digital prosperity is not metro- or elite-exclusive. Tier-2 data center-friendly policies, local start-ups, and accessible cloud access narrow the gap between high-tech development and citizen participation. Inclusion here is both social justice and economic prudence.

Institutional Adaptability

The longevity of India's regulatory framework will be based on its capacity to evolve. Agencies will have to transform from rule-making to permanent learning organizations, revising guidelines based on new technologies such as:

1. Artificial Intelligence (responsible use of data, transparency of models)
2. Quantum Computing (new encryption protocols)
3. Metaverse and IoT (collecting ambient data and consent)

To do this, MeitY and its affiliated institutions are investing in policy labs, scenario simulations, and public-private research forums. This learning curve for institutions ensures that governance is a shifting equilibrium, not a fixed framework.

The Role of International Cooperation

India's inclusion in such forums as the Global Partnership on Artificial Intelligence (GPAI), the OECD Digital Economy Policy Committee, and the G20 Digital Economy Working Group is evidence of an adherence to developing, not just following, international norms. On these forums, India promotes a "trust + growth" approach: open markets buttressed by ethical governance.

Digital diplomacy has consequently emerged as a cornerstone of foreign policy, with data-transfer arrangements and mutual-recognition frameworks being now negotiated at the same table as trade and defence.

Establishing Public Trust in Governance

Even the most advanced legislation collapses without public trust. Public outreach initiatives, awareness drives on privacy rights, grievance portals, and open disclosure standards translate formal regulation into everyday existence.

India's Digital Nagrik program seeks to teach citizens about consent, cyber hygiene, and rights to data, building people who engage in governance instead of being governed.

Economic Returns of Balanced Regulation

Balanced rule has concrete returns:

1. Investor confidence – stable policy reduces risk premiums.
2. Operational efficiency – streamlined compliance saves time-to-market.
3. Global competitiveness – compliance with global standards brings in enterprise customers.
4. Domestic innovation – SMEs and start-ups have a level playing field when norms of data-handling are clear.

The two-way interplay of regulation and enterprise thus becomes self-supporting: regulation generates trust, trust attracts investment, and investment requires better regulation.

India's Emerging Model: Regulated Openness

Through all these layers, a clear Indian model is taking shape, neither laissez-faire nor autocratic, but regulated openness. Its guiding ideas can be encapsulated as follows:

1. Sovereignty with Connectivity – Information stays in national hands but moves out through reliable conduits.
2. Compliance with Ease – Electronic gateways and single-window interfaces take the place of paper red tape.
3. Security with Transparency – Cyber protection exists alongside public transparency.
4. Growth with Responsibility – Investment offerings are linked to sustainability indices.

This balance reflects India's general democratic style: pluralistic, responsive, and practical.

The Road Ahead

As India plans to install gigawatts of new capacity, the future phase of administration will naturally be on:

1. Unified Digital Infrastructure Authority for cross-industry coordination.
2. Dynamic Data Exchange Standards facilitate interoperability between government and industry.
3. National Cyber Insurance Framework for systemic risk mitigation.
4. ESG reporting is incorporated in licensing norms.

Every step underline that economic power and moral governance cannot be separated in a data world.

From Regulation to Stewardship

As maturity comes in governance, it no longer comes across as control but as stewardship. India's journey from permissive digital growth to moral digital governance is a sign of that maturation. The nation is no longer adopting the best practices but is adding its own philosophy to the global debate: inclusivity, trust, and sovereignty can go together with innovation and enterprise.

As we transition from questions about the law to questions about logistics, the discussion changes once more, from how data is regulated to how it is energized. For each regulation drafted on paper, there will have to be a watt of electricity powering the servers that bring it into existence. The following chapter delves into this pillar of the digital world, Infrastructure, Power, and Sustainability, where technology and environment converge, and where India's desire for size must balance with its duty to the earth.

Chapter 4

Energy, Sustainability, and Design

"Energy is the golden thread that connects economic growth, social equity, and environmental sustainability." - Ban Ki-moon.

In every data center, no matter its size or purpose, there is a single concern that remains constant from the moment the first line of design is drawn. That concern is power. If the earlier chapters described data centers as the factories of the digital age, then this chapter turns toward the engines that keep those factories running, the cooling systems that keep them alive, and the ecological responsibility they must now embrace as global demand grows. Power, sustainability, and design are not parallel topics. They are the three sides of the framework on which all modern digital infrastructure rests.

Across the world, governments, companies, and citizens are asking a question that did not exist twenty years ago: What is the environmental cost of the digital convenience we enjoy? It is not an abstract question. All the photographs uploaded to the cloud, all the files stored in an enterprise server room, all the transactions traversing an encrypted channel, all the AI algorithms that take days or weeks to train, make a physical difference somewhere, manifested as electricity used, heat emitted, water vaporized, or hardware disposed of. The more reliant societies become on digital networks, the greater the need to explore how these facilities are energized and how they can develop without straining the world in which they operate.

This chapter looks at that question closely. It begins with the basic physics of running a data center, moves through the design logic that engineers employ to keep them cool and efficient, and then turns to the broader responsibility of operating them sustainably. The goal is not simply to explain how operators reduce their energy use or why cooling systems matter, but to illustrate how power and sustainability are no longer optional components. They have become central to the identity and survival of the data center industry.

The Demanding Appetite of Data Centers

From the outside, a data center may resemble a quiet structure. A room that is sealed and soundproofed. An infrastructure giving little indication of what happens inside. The reality is entirely different. Operating racks of high-density servers requires more continuous electricity than almost any other type of building. A single hyperscale facility can draw as much power as a mid-sized town. Even a colocation center serving multiple clients must maintain a level of power stability that no commercial building experiences.

The simplest way to consider the magnitude of energy used is in how the IT equipment is related to the infrastructure that supports it. All servers, however efficient, produce heat. Multiply that by hundreds or thousands of servers operating at once, and the amount of energy needed to maintain their stability is enormous. The energy draw is not limited to IT equipment alone. Power flows to cooling systems, mechanical equipment, battery banks, uninterruptible power supply units, fire-safety mechanisms, lighting, security networks, and environmental monitoring systems.

In the earliest years of data centers, power was viewed as an operational cost rather than a design principle. However, with the growth of the industry and the increasing demand for electricity, new parameters came to gauge and enhance efficiency. Of these, the PUE, or Power Usage Effectiveness, was the term that transformed the discussion.

Introduction to PUE and Efficiency Metrics

PUE is a simple ratio. It takes the total power used by a data center and compares it to the power that is used by the IT equipment. Assuming that each unit of power used by the servers in the facility demands 1.5 units of power in the facility, the PUE of the facility is 1.5. The theoretical ideal of PUE of 1.0 is always unrealistic since supporting infrastructure will never be free of power consumption. But the metric has become such a yardstick that it shows the amount of energy that is being consumed on all the other things, besides the actual computation itself.

In early enterprises, the PUE figures might be 2.0 or higher, i.e., 1 watt to power up a server, 1 watt to cool, protect, and support it. Technology was enhanced, and this made air management more accurate; the cooling systems were better, and less power distribution wastage. Large operators began reporting PUE values of 1.4, then 1.2, and in exceptional modern designs, 1.1 or even slightly lower during cooler seasons. For instance, in 2023, AWS disclosed a global fleet PUE of about 1.15, while its best site registered a PUE of 1.04.

Efficiency metrics have now expanded beyond PUE. Operators track carbon usage effectiveness, water usage effectiveness, and renewable energy percentage. Continuous airflow, server density, temperature stability, and battery cycle health measurements are monitored continuously. The wattmeter on the wall that used to be a single wall sensor has transformed into a collection of sensors that feed real-time dashboards that dictate operations.

However, even these figures are not the whole story. To enhance PUE, much complex interaction of building design, cooling architecture, rack layout, and electrical engineering is needed. It is at this point that design decisions start influencing sustainability outcomes.

Managing Heat: The Central Design Challenge

Any data center design commences on one simple fact that keeps growing: heat will build up quicker than it can be removed. As soon as a server is turned on, it heats the air in the vicinity. Repeat this by hundreds of racks and thousands of processors, and the room will become hostile unless the cooling systems come into play. The current systems do not consider the heat as a byproduct, but a design variable, which determines the airflow, equipment proximity, and energy allocation.

Air Cooling and Its Evolution

Traditional facilities relied on air-based cooling systems. Rows of Computer Room Air Conditioning units circulated chilled air either from raised floors or overhead ducts. Cold air would flow across server intakes, pick up heat, and exit from the back into the hot aisles. For many years, this approach worked, but it demanded enormous amounts of energy. As server densities increased and artificial intelligence workloads intensified, air cooling alone began to lose efficiency.

To counter this, data center architects introduced hot aisle and cold aisle containment. Rather than allowing cool and warm air to blend freely, containment systems physically separate them. Cold aisles, sealed by doors or panels, kept intake air at stable temperatures. Hot aisles confined exhaust air, allowing cooling systems to treat warm air more efficiently without fighting against thermal mixing. While data centers with traditional air-cooling report a PUE of 1.5 to 1.8 range, modern air-cooling sites reach a PUE of 1.3 to 1.5 range.

Containment alone significantly improved efficiency, but it was not the final step in the cooling story.

Liquid Cooling and the Return to Fundamentals

At high densities, air becomes a less efficient medium for heat transfer. Liquids, by contrast, carry heat more effectively. This reality pushed the industry toward liquid cooling technologies, which bring coolant closer to the heat source.

Two major variations have become common:

Direct to chip cooling - In this process, cold plates pass liquid over processors.

Immersion cooling - Here, the servers are flooded in a non-conductive fluid that is able to absorb heat more effectively than air.

Facilities that support artificial intelligence, machine learning, and high-performance computing have paid special attention to immersion cooling. These loads produce excessive heat that cannot be handled by conventional air systems without a lot of energy consumption. However, in immersion systems, heat dissipation and absorption occur with minimum mechanical work, which consumes very little energy. Liquid cooling systems show 10-20% improvement in whole-site PUE levels, which helps sites towards 1.3 to 1.2 PUE.

Heat Reuse: Turning Waste into Value

A major shift in data center thinking has been the recognition that the heat they produce need not be wasted. In colder climates, heated air or water from a data center can be fed into district heating systems, warming residential buildings, offices, schools, and other facilities. Several European cities, including Stockholm, have implemented heat reuse networks where data centers contribute to the community's heating requirements.

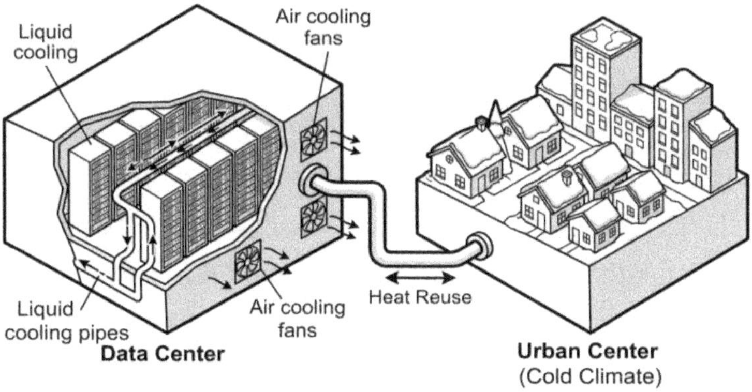

Figure 4.1 Heat Reuse System

Heat reuse presents attractive possibilities in urban settings, where thermal energy has immediate demand. Instead of cooling towers releasing waste heat into the atmosphere, that energy becomes a local resource. Not every climate or location can adopt heat reuse easily, but where feasible, it transforms data centers from energy consumers into contributors.

This progression from air cooling to liquid cooling and eventually heat reuse shows how engineering choices influence sustainability. Yet cooling is only one part of the equation. The other, increasingly central aspect, is the source of power itself.

The Shift Toward Renewable Energy

"Sustainability is no longer about doing less harm. It's about doing more good." - Jochen Zeitz.

A decade ago, renewable power was considered aspirational. Solar and wind were viewed as contributors, not cornerstones. That perception has changed. Nowadays, the data center operators are the large operators that include a significant portion of renewable energy in their operational concepts. Long-term energy plans are now based on solar parks, wind farms, and green power purchase agreements.

They are not symbolic gestures. They are financial needs and working obligations.

Megabusters and especially hyperscalers have led by example. In the United States, Google data centers have embraced contracts of renewable power that are equivalent to or more than their yearly usage. Operators in Hyderabad and Mumbai, India, have started to incorporate solar energy into the supply, offsetting grid energy with a dedicated off-site generation. There are two benefits of renewable energy: it will minimize the carbon footprint, and the energy costs will be volatile in the long term.

But there is the issue of renewable power with engineering problems. Solar energy is dependent on the weather and time of day. Wind output varies by season. In order to deal with such variances, contemporary data centers have instituted hybrid energy plans that integrate renewables and grid power, backup batteries, and high-efficiency generators. It can be concluded that energy storage systems are currently in a critical role, which balances the intermittent quality of renewables and provides constant operation.

Such a transition of facilities that depend on fossil fuels to renewable ones is the first step in a greater shift in the sustainability of data centers. Nevertheless, the environmental responsibility goes beyond that.

Although renewable energy and efficiency indicators are redefining the operation of data centers, there is a bigger question. How much environmental impact would this industry have overall, and what can be done to decrease it without taking away reliability? The issue cannot be answered only through the lens of power consumption. A data center's environmental impact extends to everything it touches: water, land, heat, air, and hardware. If the previous part introduced why power and cooling strategies have taken center stage, the next step is to follow those strategies into the larger ecological landscape and see how they interact with global sustainability expectations.

The Environmental Footprint of a Digital Society

Digital systems do not have a weight, but the infrastructure does. The physical requirements of data centers are increasing as more workloads move to the cloud and the number of artificial intelligence operations increases in more places across the globe. Their environmental impact can be perceived in three general aspects, which are carbon emissions, water consumption, and hardware waste.

Carbon Emissions: The Challenge Behind Invisible Activity

Today, data centers are responsible for roughly 0.5–1% of energy-related CO_2 emissions globally. Most of a data center's carbon footprint originates from the electricity it uses. In regions where power grids still rely heavily on fossil fuels, a facility's indirect emissions can be substantial. Even when efficiency is optimized at the facility level, the source of the energy determines its true environmental cost.

This is why renewable energy adoption is more than a design preference. It is the most direct path toward reducing carbon emissions. Operators today measure carbon not only through annual totals but through hourly or regional intensity. Some advanced facilities adjust workloads dynamically based on renewable availability, shifting non-urgent processes to periods when solar or wind supply is high. This practice, often described as carbon-aware computing, aligns operational demand with the natural rhythms of renewable production.

Yet carbon emissions are not determined solely by electricity usage. Every part of a data center's life cycle, from construction materials to backup systems, contributes to its carbon profile. Manufacture of concrete, production of steel, battery manufacturing, and transportation all have embodied carbon. With the evolution of sustainability standards, operators are currently mandated to consider operational carbon (which comes with the continued consumption of electricity) and embodied carbon (which comes with the building and machinery).

Embodied carbon reduction requires proper planning. Engineers are now opting to use low-carbon cement replacement, recycled steel, and energy-efficient processes of production. Other companies have also started using old buildings rather than constructing new ones, as they generate less carbon footprint compared to new construction. These steps differ depending on the regions and availability of the same; however, it is a change in thought. Sustainable design begins long before the first server is installed.

Water Consumption: The Cooling Secret of Dependency.

Water sustains most of the systems, which maintain a data center cool in case electricity is keeping it running. Conventional cooling towers rely on the evaporative cooling mechanism, which may use large amounts of water. The International Energy Agency estimates that data centers worldwide consume more than about 560 billion litres of water annually. This reliance is problematic in areas where there is a water shortage.

In order to solve this, the latest data centers are testing various methods:

- Recycled closed-loop cooling systems that use the same water over and over, as opposed to constant evaporation.
- Air-cooled chillers eliminate or minimize water consumption but consume more electricity.

- Hybrid systems that change the cooling medium to air or water depending on the environment.
- During the good seasons, rainwater harvesting is used for cooling or humidifying.

The most appropriate solution is related to the climate and geography. Air-based or hybrid cooling would be more realistic in arid areas where the pace of evaporation is high. In lower temperatures, outside air is cooled by methods that do not demand a lot of water since large masses of water are not essential. Water Usability Effectiveness (WUE) is now measured together with PUE to understand consumption trends and to design long-term solutions. A typical data center's WUE is averaged at 1.8 liters per kWh of IT energy. Leading operators report WUE values as low as about 0.2 L/kWh.

Wastewater management is also a part of water stewardship. Treatment of cooling circuits and cleaning cycles should be done with accuracy so that contamination does not occur. In highly populated industrial factories, there are data centers that have started to use treated wastewater from municipal plants instead of fresh groundwater. This will help minimize the pressure of demands on the local supplies and will help synchronize the expansion of data centers with the requirements of the whole community.

E-Waste: The Material History of the Digital World.

Every server has a lifespan. The storage devices eventually become worn out. Networking equipment goes outdated. Electronic waste is bound to occur because of the fast rate at which technological equipment is developed, and consequently, hardware is also changing hands. One of the most rapidly increasing waste streams in the world is e-waste; data centers are also contributors to it.

Operators have two ways to address e-waste: reduction and responsible disposal.

Reduction entails ensuring, where possible, that equipment life is extended by reconfiguration, virtualization, and component-level upgrades. Numerous hyperscale operators are adhering to the rules of a circular economy where the decommissioned equipment is repurposed and reused in the same facility or resold to smaller operators. The reusable components are removed, e.g., a memory module or power supply, and used again.

Proper disposal requires high compliance with the environmental standards. The recycling centers are certified to extract value-added metals like copper, gold, and silver without allowing toxic wastes like lead, mercury, and some plastics to find their way into landfills. Other manufacturers currently manufacture servers with modular parts that can be easily recycled, and less hazardous waste is produced.

The e-waste management is a sustainable management and not merely a matter of operation. It is a method to attain a process of looping the resource requirements of the digital infrastructure.

Designing Data Centers to be Sustainable: A New Paradigm

Environmental responsibility is one of the pressures that has influenced the approach that architects and engineers take when designing a data center. A building that used to be a plain power and cooling infrastructure has now transformed itself into a well-calculated construction that is responsive to energy realities, climatic factors, land restraints, and regulatory demands.

There are four principles of sustainable data center design:

1. Efficiency by Architecture

Efficient designs start with a building layout. The server rooms will be designed in a way that maximises airflow, minimises pressure differentials, and makes the most efficient use of cooling. Energy consumption depends on the orientation of the building, insulation quality, as well as the location of the mechanical rooms.

Computational fluid dynamics (CFD) models are used in models that represent airflow before the start of the construction to enable a designer to know the hot spots and rectify the airflow problem early. The choice of raised floors, overhead containment, and aisle design is highly accurate so as to manage the thermal behaviour.

2. Elasticity of Future Workloads

The higher rack densities needed by artificial intelligence, high-performance computers, and edge workloads than by conventional enterprise applications. Sustainable structures have modular power and cooling areas that can be expanded in the future without necessarily redesigning the total building.

Scalable power distribution, flexible cooling channels, and modular containment are the design features of flexible designs. This decreases the amount of material footprint of future upgrades and precludes unwarranted reconstructions.

3. Connection with Local Environment

The idea of sustainable design is geographically based. Free cooling opportunities are enjoyed in data centers in colder regions. In places located along the coast, the cooling of seawater can be investigated, and in such cases, the environmental regulations permit it. Cities can incorporate heat reuse, whereby warm water or air is distributed to the adjacent residential or industrial areas.

Environmental impact assessment is now being conducted at the initial phase of the planning by designers to ensure that the facility is in sync with the ecological conditions in the region.

4. Built for Longevity

As data centers contain high embodied carbon, the most eco-friendly buildings are ones that are built to endure decades. The heavy-duty structural elements, robust power distribution, and adjustable cooling facilities minimize the necessity of reconstruction. The long-term facilities reduce the life cycle environmental impact.

All these principles constitute sustainable data centers. They signify a change of perception towards such facilities being high consumption buildings to being long-term assets that have to strike a balance between digital demand and environmental stewardship.

New Concepts of Sustainable Operations

A great number of operators are investigating methodologies that lie between engineering and ecological mindset. Among them are:

- AI-generated energy optimization which optimizes the cooling system and power distribution in real time.
- Dynamic thermal limits, which enable servers to work at relatively high temperatures, minimise cooling load.

- Local renewable generation, solar rooftops, and microgrids.
- Green building certification, by LEED, IGBCor national alternatives.
- The energy storage system, which involves batteries, eliminates the use of diesel generators.

These practices transform sustainability into a concept into a discipline of operation.

Real World Case Studies

The theory of sustainable data center design can be made more real with reference to real examples. Some of the projects around the globe show how renewable energy, efficient cooling, and progressive architecture can converge.

Throughout the globe, there are few projects where the industry has made reference bases not just due to the display of technical expertise but also due to the way they express alternative methods of dealing with the same challenges. These case studies demonstrate that sustainability cannot be a one-answer solution being replicated everywhere. Rather, it is geographically, climatically, infrastructurally, and communally adaptive.

Renewable Energy-Powered Data Centers: Lessons from Google, USA

Few companies have shaped the global conversation around sustainable digital infrastructure as directly as

Google. What began as corporate environmental responsibility has grown into one of the largest renewable energy portfolios in the world, tied closely to its network of data centers. Google's facilities across the United States illustrate how design, procurement strategy, and operational discipline work together to reduce carbon footprints in meaningful ways.

A Long-Term Approach to Renewable Integration

More than a decade ago, Google began signing power purchase agreements with renewable developers, locking in wind and solar supplies for long periods. These contracts created predictable energy costs while ensuring that an equivalent amount of renewable energy was added to the grids supporting their facilities. The company's data centers in Iowa, Oklahoma, and Oregon particularly benefited from this strategy because those regions had strong wind energy resources.

The outcome was twofold. First, Google was able to match nearly all of its annual energy consumption with renewable production, achieving a milestone that set a precedent for the entire cloud industry. Second, it demonstrated that large-scale digital facilities could integrate renewable energy without compromising reliability.

Operational Efficiency to Match Renewable Goals

Alongside renewable procurement, Google invested heavily in improving PUE across its data centers. Through refined air management, advanced liquid cooling approaches in selected high-density zones, and intelligent power distribution, the company's facilities achieved PUE values much lower than industry averages. Machine learning models were later introduced to predict cooling requirements and tune equipment in real time. These models reduced energy consumption without disturbing server stability.

Local Impact and Ecosystem Influence

Google's progress shaped broader industry norms. By openly publishing their methods and annual sustainability results, the company encouraged others to treat renewable integration as an achievable operational goal rather than a symbolic gesture. Their model also influenced policy discussions on renewable adoption, especially in states where large consumers, including data centers, were negotiating for cleaner electricity options.

Relevance to Emerging Markets

For regions such as India, where renewable availability varies across states, Google's strategy offers a roadmap. It shows that long-term renewable contracts, combined with efficient facility design, can reduce environmental impact while supporting the high-demand workloads of modern cloud computing.

Campuses Powering Up Renewably in India: Hyderabad and Mumbai

The data center market in India is expanding at a very high rate, and sustainability has been a key issue since new data centers are joining the market. Hyderabad and Mumbai have both become strategic in terms of their connectivity and industrial ecosystems, as well as in the fact that operators operating within them have started to consider renewable energy when planning and designing their operations.

Hyderabad: New Green Ambivalent Cluster

The developments in data centers in Hyderabad are characterized by a combination of renewable sourcing and wise design. There are solar farms in Telangana that operators in the region have contracted for the power, and therefore, a part of their load could be supplied by clean energy. It is also through campus higher outdoor temperatures that several campuses use them as a chance to investigate other cooling options that include chilled-water systems and hybrid cooling designs that alternate between water and air based on the time of the day.

Local policy assistance has also contributed. The fact that the state is open to renewable integration and has special substations in the area of data center clusters makes the planning of power more predictable. Consequently, the new campuses of Hyderabad are suitable proof of how in India, a significant part of digital infrastructure can be created, not entirely reliant on fossil-intensive power grids.

Mumbai: Density vs Sustainability

The City of Mumbai offers another form of challenges to sustainability due to its geographical location, where it is bounded by the sea and the land. The data centers of the city should be able to support high rack densities in small footprints and contain heat and power very precisely. The operators have reacted by putting money into cooling systems that consume less energy, such as advanced air containment and liquid-assisted cooling systems.

In Mumbai, solar uptake is normally through off-site power generation. Due to the small size of the area on the roof, campuses collaborate with solar farms in Maharashtra and the surrounding states to achieve their renewable goals. Other plants are trying out heat recovery alternatives, although urban planning limits the reuse of the heat scale. Rather, efficiency, the benefit of better PUE, and best cooling performance have been in the limelight.

Moving the National Conversation

Such projects in Hyderabad and Mumbai show that sustainable data centers in India must have a regional strategy. The techniques are dependent on climate, maturity of the power grid, land availability, and local policy. The developments observed in the two cities show that the sustainability situation in India will not be achieved by any one model but a set of region-specific approaches that embody the local realities.

Stockholm Heat Reuse Model: A Circular Approach to Sustainability

---

"Innovation is seeing what everybody has seen and thinking what nobody has thought." - Albert Szent-Györgyi

---

The use of the heat reuse case study in Stockholm is one of the most effective and simplest examples. The city takes waste heat as a resource instead of seeing it as a problem. Because Stockholm experiences long periods of cold weather, heat generated by servers becomes valuable for residential and commercial heating systems.

District Heating Meets Digital Infrastructure

Stockholm's district heating network is one of the most extensive in Europe. When data centers began exploring sustainable options, collaboration with energy companies emerged naturally. Instead of relying solely on cooling towers to expel heat, data centers across the region now use heat exchangers to transfer thermal energy into the district heating grid.

This process has several advantages:

- It reduces the need for cooling equipment to work as intensively.
- It lowers the carbon footprint of data centers by minimizing energy loss.
- It supplies clean, stable heat to thousands of homes.

For the city, this model helps balance energy systems and reduces the need for fossil-based heating sources during winter.

Retrofitting for the Future

The most compelling aspect of Stockholm's model is that it has been successful even in retrofitted facilities. Older data centers were equipped with redesigned heat pipelines and integrated into the district heating network through coordinated engineering.

A Model for Cold-Climate Cities

Stockholm's experience is not easily transferable to warmer regions, where heat reuse may not have immediate demand. Yet, it provides an important lesson: sustainable design gains strength from local conditions. Climate-responsive solutions can multiply environmental benefits when cities and operators collaborate.

Underwater Data Centers: China's Hilander and U.S. Subsea Concepts

An alternative that has been more experimental in regard to data center sustainability has been the idea of laying facilities underwater. In their infancy, these projects demonstrate that creative thinking can result in surprising answers to the problems of the long-term scope.

China's Hilander Initiative

China's Hilander underwater data center concept uses the thermal stability of deep water to cool servers more efficiently. The seawater itself has low and more uniform temperatures when compared to the air at the surface and thus consumes less mechanical cooling. The building is essentially a closed container that accommodates racks of servers, anchored at the sea and linked to land-based power and network.

Several benefits can be noted:

- Lower cooling energy requirements.
- Reduced land use in dense coastal regions.
- Stable thermal environment year-round.

While the project is still evolving, it demonstrates a direction in which future data center designs might move, particularly in nations with limited land availability.

U.S. Subsea Experiments

In the United States, subsea data center experiments have explored similar ideas, focusing on the potential energy savings associated with underwater deployment and the ease of maintaining modular sealed units. Past experiments have demonstrated that submersible structures have inherent high heat transfer and can be installed quite easily in a comparatively short time as compared to conventional plants.

Such projects also pose new concerns regarding long-term maintenance, environmental effects on marine ecosystems, and the economic viability of large-scale underwater implementation. However, they also demonstrate that sustainability issues tend to stimulate experimentation on the boundary.

The Underwater Model Interpretation

The data centers placed underwater will hardly be able to substitute those installed on the ground in the near future, but they present an interesting picture of flexibility in design. They emphasize the fact that the digital infrastructure does not have to be based on a traditional building only and can be adapted to the environment, having the benefits of natural cooling.

The examples of the United States, India, Europe, and recent experiments of the subsea constitute a varied topography of approach, each of which is contextually oriented. But these differences have a common thread. With the growth of digital systems being used in the global world, the accountability of these systems increases equally. Power decisions, design rationale, and environmental responsibility have ceased to be a peripheral worry but are an element of the data center planning. This change is indicative of a wider change in the way societies perceive digital infrastructure, not as a single technical object but as a component of a broader environmental system that supports contemporary life.

What These Case Studies Reveal About the Future of Data Center Design

When observed together, these projects outline a steady movement toward integrated thinking. Earlier generations of data centers were designed primarily around reliability and cost. Sustainability, while acknowledged, did not shape architectural decisions in a significant way. That division has eroded. Today, sustainability is inseparable from design, operations, and long-term planning.

Several patterns emerge clearly.

1. Energy Strategies Are Becoming Long-Horizon Commitments

Companies no longer view energy procurement as an annual operational task. Instead, they approach it as a decades-long relationship involving renewable contracts, storage planning, and coordination with local utilities. Google's renewable agreements, India's state-supported green corridors, and the steady expansion of wind and solar near data center clusters all reflect this long-horizon approach.

Energy strategy now shapes where facilities are built, how they are engineered, and how workloads are scheduled. In time, energy availability may determine the geography of global computing, just as access to ports once guided the industrial economy.

2. Cooling Systems Are Evolving Faster Than Any Other Component

From containment systems to liquid cooling, seawater cooling, and even the thermal stability of underwater environments, cooling technology has become the most dynamic area of data center design. The need stems directly from new workloads. Artificial intelligence and high-performance computing generate heat profiles beyond those of conventional enterprise applications.

The cooling innovation rate implies that the industry is getting ready to accept workloads that would be more challenging. Immersion cooling, formerly viewed as an experimental technique, is now potentially practical in large-scale applications. Thermal management is turning from a building-level issue into a rack-level and even chip-level engineering issue.

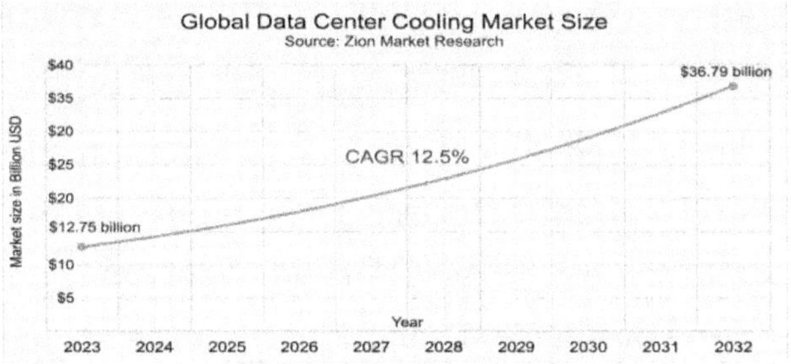

Figure 4.2 Cooling Market Size

3. Environmental Integration Is Much Localized

The heat reuse in Stockholm is due to the fact that the city has a developed district heating network, facilitated by the cold weather needs. The application of hybrid cooling in Hyderabad is due to the hot climate and power policy in the area. Experiments in underwater research in China and the U.S. are based on the availability of coastal waters and closeness to research facilities.

These variations highlight an important fact that sustainable design is most effective in its response to local realities. There is no standardized blueprint. Rather, the industry is shifting to climate-sensitive planning in which the design choices are organically a result of the environment within which the facility is situated.

4. Operators Are Looking at the Lifecycle of Their Footprint

Efficiency used to be solely about performance in the operations. Nowadays, focus is given throughout the lifecycle of a data center: its construction, operation, equipment refresh, and decommissioning. This has pushed operators to track embodied carbon, reduce water dependency, and plan for responsible hardware retirement.

These steps signal a maturing industry, one that understands that environmental impact cannot be minimized through operational changes alone. Meaningful sustainability requires choices that extend across decades.

Balancing Growth with Responsibility

The digital economy continues to expand. Cloud adoption rises annually, artificial intelligence requires exponentially more computation, and the number of connected devices grows without pause. The result is clear: data center capacity will continue to increase worldwide. With that growth comes an essential responsibility. Operators must ensure that efficiency gains and renewable integration keep pace with demand.

There is no contradiction between digital expansion and environmental stewardship. What matters is the rate at which sustainability advancements rise alongside technological progress. In regions where energy grids are transitioning slowly, data centers can accelerate the shift by committing early to renewable procurement. In water-stressed areas, cooling innovation can reduce dependency on local supplies. In colder climates, heat reuse can convert waste energy into community benefit.

The industry's trajectory shows that sustainability does not oppose growth; it strengthens it. Efficient facilities reduce operational costs. Renewable energy stabilizes long-term expenditure. Heat reuse forges new partnerships between operators and communities. Design improvements allow higher densities in smaller footprints, reducing land pressure. Every sustainable decision creates a structural advantage that supports future development.

Sustainability as an Element of Trust

Across all these cases, another theme emerges quietly but powerfully. Data centers have become public infrastructure in everything but name. They enable services that citizens, companies, and governments rely on daily. With that role comes scrutiny. The use of energy, water resources, and carbon emissions is no longer a secret. They are the objects of social attention, politics, and administrative control.

Companies that are sustainable gain credibility within their areas of operation. This trust is not abstract. It shows itself in the form of less contentious regulatory concessions, enhanced local alliances, and more robust long-term operations. On the other hand, the ones that do not integrate environmental planning in the digital infrastructure should expect resistance from the people, restrictions in policies, and slow growth.

The Stockholm retrofit project is one such project that has occurred in a positive collaboration. The operators and residents benefit by matching the heating requirements of the city with the data centers. However, as opposed to that, the countries where moratoriums on the construction of the data center were imposed learned that the growth without environmental balancing can be stopped lightning fast.

Sustainability has hence become a strategic need, rather than a moral need. Stability in an industry relies on the confidence of the people, as a key asset.

Design Priorities Are Shifting Permanently

As new data centers are planned, especially in emerging markets, designers and operators are giving more weight to the environmental dimension than any previous generation of infrastructure. Several priorities are beginning to define this shift:

- Energy efficiency is now a baseline expectation, not a differentiator.
- Renewable integration is becoming a long-term financial strategy.
- Water stewardship is emerging as a central criterion in site selection.

- Circular economy principles are influencing hardware planning.
- The city data centers are not just judged by size but also on how they can help the community.

This development is an indication that the sector is going through a bigger transformation. The uptime or capacity is no longer the only criterion for judging data centers, but also how they work within ecological constraints.

As discussions around data centers shift from cost to green oversight, we see power, cooling, and design emerging as key components in shaping societal perceptions of the digital world. Innovations in efficiency and renewable energy highlight the importance of the relationship between technology and nature. The future of data center facilities will demand designers who think long-term, engineers who understand hardware and climate trends, and operators who balance reliability with resource stewardship, creating a cohesive infrastructure that transcends individual elements.

It is at the crossroads of all these areas that design stands. Thoughtful architecture helps reduce energy waste. Scalable layouts allow facilities to adjust to new workloads without rebuilding. Strategic material choices lower embodied carbon over time. These details influence the long-term environmental footprint even before the first server is turned on.

There is growing adoption of data center design as a field that combines engineering with environmental science. Strong and modern facilities are not the best, but flexible, efficient, and responsive to the environment they are located in. With the increasing centrality of the digital infrastructure to economic and social life, design will dictate whether the infrastructure can usher in an advanced era without an excessive commitment to natural systems.

The Changing Expectations Around Power

Data centers have always relied on electricity, yet there is an increased expectation on what data centers can demand of the electricity. Reliability is always required, but no longer suffices. The question of the source of electricity, its carbon intensity, stability with time, and responsibility of use within the facility are issues that stakeholders are now concerned with. It has become a cause of conscious consumption rather than raw consumption.

This change has a practical implication. Renewable integration involves long-term contracts. Close collaboration is required for efficiency between the IT and facilities teams. Grid stability necessitates an interest in battery storage, microgrids, and more intelligent energy management systems. These decisions influence the carbon footprint of a particular location and the energy ecosystem of an area where they are implemented.

Cooling as an Innovation Force

One of the most dynamic spheres of technological change has become cooling. The traditional systems are being changed into more accurate, more climate-receptive practices that can easily adapt to the fluctuation in workload and weather. Regardless of whether a given facility is air-based, liquid-based, seawater-based, or has other advantages of natural cooling that come with underwater deployment, cooling technology is coming out as the balance between environmental responsibility and operational performance.

The progress that has been achieved over the past ten years indicates that cooling will continue to be a source of innovation in the future. With the growing need for processing, the industry will keep experimenting with novel combinations of materials, fluids, and architectural designs that can handle heat without increasing the use of resources. This search for balance is likely to define the engineering culture of the sector, just as fault tolerance once defined the early years of data center design.

Sustainability as a Collective Correspondence

Sustainability has transitioned from being an external consideration to a core element of data center planning, affecting aspects such as site selection, energy sourcing, cooling methods, and equipment lifecycle management.

Examples like Stockholm's advancements, renewable energy commitments in the United States and India, and underwater experimental systems showcase an evolving perspective on sustainable infrastructure. The effectiveness of heat reuse in cold climates, solar integration in sunny regions, and the cooling advantages of coastal waters highlight how varied approaches can coexist harmoniously. This diversity reflects the industry's maturity, as it adapts to local conditions while embracing its global responsibility to realize sustainable potential.

Although technology firms and service providers are the ones who are best positioned to ensure that data centers are sustainable, the problem goes way beyond companies. Policy outcomes are manipulated by governments, which control the electricity policies, the urban planning, and the environmental policies. The utilities determine the stability and cleanliness of the local power grids. Embodied carbon is an effect of production techniques by the manufacturers. Consumers also play an indirect role in their consumption through the services they take and the environmental quality that they demand.

Sustainability has thus become a common effort. As data centers evolve, the responsibility for guiding them toward cleaner, more efficient futures rests with communities, policymakers, engineers, and end-users alike. The future of digital infrastructure will be shaped not just by technical decisions but by the values societies place on environmental stewardship.

A Transition Toward Risk, Reliability, and People

The themes of this chapter lead naturally to the concerns of the next. Power, sustainability, and design define how datacenters interact with the environment. The next set of questions deals with how these facilities interact with disruption, uncertainty, and the people who keep them running.

Digital infrastructure must endure more than resource constraints. It must withstand natural disasters, supply chain instability, cyber threats, physical security concerns, and the constant pressure for uninterrupted availability. Behind every facility is a human workforce trained to manage systems that cannot afford to fail. Their expertise, judgment, and discipline form the final layer of reliability in a landscape full of complex systems.

Chapter 5 will explore these dimensions. It will examine the risks that data centers face, how they prepare for them, and the skills required to sustain their operations. As sustainability defines their relationship with the environment, resilience defines their relationship with the world's uncertainties

Chapter 5

Risks, Reliability, and Human Capital

"The greatest danger in complex systems is not failure itself, but the belief that failure has been eliminated." - Richard Cook

The Burden of Being Always On

There is a quiet promise embedded in modern life, one that most people never consciously acknowledge. It is the promise that systems will work. That payments will go through. That messages will arrive. That data will be there when needed. This promise is rarely articulated, yet its violation is immediately felt.

When a banking application freezes, anxiety follows. When a logistics platform goes offline, supply chains stall. When healthcare records become inaccessible, consequences move beyond inconvenience into harm. In a digital economy, interruption is not merely technical. It is social, economic, and sometimes political.

Data centers are not monuments to innovation, nor symbols of technological ambition. They are, at their core, institutions of reliability. Their central obligation is not speed or novelty, but continuity. To remain operational when surrounding systems fail. To absorb shocks without transmitting them outward. To stay unremarkable, even under extraordinary strain. This task is far more demanding than it looks.

Reliability is an Expectation. It is not a Feature

In earlier eras of computing, downtime was tolerated. Systems were local, limited in scope, and understood to be imperfect. Maintenance windows were announced. Services paused. Users adjusted. Failure, while inconvenient, was accepted as part of the bargain.

That social contract has dissolved.

Today's digital infrastructure supports continuous commerce. It looks forward to real-time communication. It offers critical public services. There is no agreed-upon pause. Global platforms operate across time zones. Financial markets never truly close. Emergency services cannot wait for scheduled maintenance.

In this environment, reliability ceases to be a technical attribute and becomes a societal expectation.

This shift places an asymmetric burden on data centers. When they succeed, nothing happens. When they fail, everything does.

Understanding Uptime Beyond Percentages

Reliability is often expressed numerically. Uptime percentages are quoted with confidence. As if numbers alone can convey assurance. Yet these figures conceal more than they reveal.

A system designed for 99.9 percent uptime may be unavailable for nearly nine hours each year. At 99.99 percent, downtime falls to under an hour. At 99.999 percent, often referred to as "five nines," the margin narrows to minutes.

On paper, these differences seem marginal. In practice, they define entirely different worlds of consequence.

For a social media platform, minutes of downtime may generate frustration and public criticism. For a stock exchange, minutes can trigger market instability. For a hospital network, minutes can place lives at risk. The same percentage, applied across contexts, carries radically different weight.

This is why reliability cannot be discussed abstractly. It must be understood in relation to impact.

The U.S. Department of Defence has defined reliability formally in its reliability standards. Reliability is the ability to perform a required function under stated conditions for a specified period of time. This definition looks clinical. But it captures a critical truth. Reliability exists only within context. However, when it comes to conditions, duration, and purpose, it's inseparable.

Designing for Failure, Not for Perfection

Here's one of the most persistent misconceptions about high-availability systems. They are built to prevent failure. The reality is much different. These systems are built with the assumption that failure will happen.

Power grids fail. Cooling equipment degrades. Network links are severed. Software behaves unpredictably. Human beings make mistakes. The question is never whether something will go wrong. It is whether the system can continue functioning when it does.

This philosophy governs modern data center design.

Power infrastructure is never singular. Facilities draw electricity from multiple substations, often through independent transmission paths. Transformers are duplicated. Switchgear is segmented. Uninterruptible power supply systems sit between grid power and IT load. They absorb fluctuations and bridge short interruptions. Diesel or gas generators stand ready, tested weekly, capable of carrying full load for extended periods.

Cooling systems follow the same logic. Chillers are installed in excess of peak demand. Pumps are duplicated. Airflow paths are isolated to prevent cascading heat buildup. In high-density environments, liquid cooling loops are compartmentalized so that a localized issue does not spread.

Network architecture is similarly layered. Multiple carriers enter the facility through physically separated routes. Core switches are mirrored. Routing protocols are designed to fail over automatically, without human intervention. In many facilities, network components fail daily without a user-visible impact.

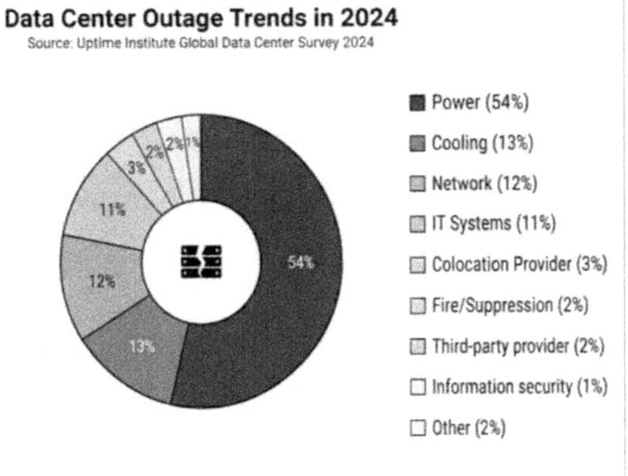

Figure 5.1 Data Center Outage Trend

This is not redundancy as insurance. It is a redundancy as a normal operation.

The Tier System as a Language of Risk

To make reliability concrete, the data center industry uses tier classifications, most notably those defined by the Uptime Institute. These tiers describe how much interruption a facility can tolerate and under what conditions.

At a high level, the progression is straightforward.

1. Tier I support basic operations but requires planned downtime for maintenance.
2. Tier II adds redundant components, reducing risk but still needing shutdowns for certain work.
3. Tier III facilities are concurrently maintainable, meaning any single component can be taken out of service without stopping operations.
4. Tier IV systems are fully fault tolerant, designed to absorb the failure of any one component without disruption.

The tiers themselves are not a badge of superiority. They are a shorthand for risk appetite. What matters is whether the chosen tier matches the role the system plays.

A small internal application can run perfectly well in a Tier II environment. A national payment network cannot. The mistake is not selecting a lower tier. The mistake is underestimating what failure would actually cost.

Reliability, in this sense, is a business decision expressed through engineering.

TIA-942: An Alternative Standards Lens

Alongside the Uptime Institute framework, another widely used standard is **TIA-942**, developed by the Telecommunications Industry Association. While the intent is similar, TIA-942 approaches reliability through a more infrastructure-centric lens.

TIA-942 defines **Rated levels** (Rated 1 through Rated 4) that broadly align with Tier I through Tier IV, but with a stronger emphasis on specific design and construction requirements. These include detailed guidance on cabling architecture, power distribution, grounding, cooling, fire protection, and physical security.

In brief:

- **Rated 1** focuses on basic site infrastructure with minimal redundancy.
- **Rated 2** introduces redundant capacity components, improving resilience but still allowing planned downtime.

- **Rated 3** requires concurrently maintainable systems across power, cooling, and cabling.
- **Rated 4** mandates fault tolerance, with complete redundancy and isolation to prevent single points of failure.

Where Uptime tiers describe outcomes, TIA-942 prescribes how those outcomes should be achieved at a physical and architectural level. Many organizations use the two together: Uptime tiers to frame availability goals, and TIA-942 to guide detailed design and implementation.

Both standards ultimately serve the same purpose. They give organizations a shared language to discuss risk, investment, and consequence, long before failure forces the conversation.

Operational Discipline and the Human Factor

Even the most robustly designed facility can be undone by poor operations. History is unambiguous on this point. The majority of major outages are not caused by equipment failure, but by human error.

A misconfigured switch. A mistimed maintenance activity. A misunderstood alarm. These are small actions with large consequences.

To mitigate this, high-reliability data centers operate under strict procedural discipline. Change management processes govern every modification, no matter how minor. Tasks are documented. Approvals are layered. Rollback plans are mandatory. Maintenance is scheduled during low-risk windows, even when systems are designed to remain live.

Control rooms operate continuously; monitoring thousands of data points related to power quality, thermal conditions, network performance, and physical security. Alerts are prioritized. Escalation paths are defined. Silence is treated with suspicion.

Technicians are trained not only in systems but in behaviour. Checklists exist not because staff lack expertise, but because expertise degrades under pressure. In environments where a single mistake can cascade across millions of users, process becomes protection. Multiple standards like TIA-942 (operations aspects), Uptime Institute M&O criteria, ISO 27001, ITIL/ISO 20000 provide frameworks to human factor improvements in data center operations.

Reliability as a Commercial Contract

From the customer's perspective, reliability is often invisible until it fails. It is experienced as a guarantee, embedded in service level agreements that promise availability, response times, and remedies.

For operators, these guarantees carry financial and reputational weight. Penalties for downtime can be severe. More damaging still is the erosion of trust that follows visible failure.

To offer such guarantees, operators must invest ahead of demand. They must build capacity that may sit idle for years. They must staff for peak scenarios that rarely occur. They must absorb costs that are not immediately recoverable.

In doing so, they transform uncertainty into predictability for their customers.

This transformation is the quiet economic function of data centers. They allow businesses, governments, and institutions to operate as if infrastructure were infinite, even though it is anything but.

Systemic Reliability in a Connected World

Modern digital services do not reside within single facilities. They span regions, continents, and jurisdictions. Applications rely on external services. Data is replicated continuously. Traffic is routed dynamically.

This interconnectedness has redefined reliability.

A single data center may operate flawlessly while a service degrades elsewhere. Latency increases. Dependencies fail. Users experience disruption without a clear point of failure.

Reliability, therefore, must be systemic. It must account for interactions across facilities, networks, software layers, and organizational boundaries. Testing must extend beyond individual sites to entire service architectures. Failover must be rehearsed not just locally, but globally.

This complexity introduces new fragility, but also new opportunity. Systems that fail gracefully at scale can achieve resilience that isolated systems never could.

The Cost of Being Invisible

Perhaps the most challenging aspect of reliability is its invisibility. Success is silent. Failure is public.

A data center may operate flawlessly for years, only to be remembered for a single incident. Media narratives form quickly. Public confidence erodes faster than technical issues can be resolved. The asymmetry is unforgiving.

This reality shapes the culture of high-reliability operations. Pride is muted. Vigilance is constant. The goal is not recognition, but the absence of attention.

In this context, reliability is not an outcome. It is a burden carried continuously, often without acknowledgment.

The Architecture of Threat

Risk, in the context of data centers, is not an occasional disruption. It is a permanent condition. Every operational day unfolds within an environment shaped by threats that are technical, human, economic, and geopolitical, often all at once.

What makes these threats especially difficult is that they do not exist in isolation. They overlap. They amplify one another. And they frequently remain invisible until their effects are already unfolding.

Complexity as the First Risk

Modern data centers are among the most complex engineered systems in existence. They combine electrical grids, mechanical systems, software-defined networks, virtualization layers, automation platforms, and human decision-making into a single operational organism.

This complexity is not accidental. It is the price of scale and efficiency. But it comes with an inherent cost.

As sociologist Charles Perrow observed in his work on system failures, “Complex systems are intrinsically hazardous systems.” His argument was not that complexity should be avoided, but that it inevitably produces failure modes that cannot be fully anticipated.

Data centers exemplify this principle. A configuration change intended to improve performance may interact unexpectedly with a security rule. An automated failover mechanism may respond correctly to one failure while triggering another elsewhere. Systems designed to self-heal may do so faster than operators can observe, masking root causes until they resurface under heavier load.

Complexity, in other words, creates risk not through negligence, but through interaction.

Cybersecurity: The Expanding Attack Surface

Cybersecurity risk has grown alongside data center scale. What was once a matter of perimeter defence has become a challenge of continuous exposure across layers that extend far beyond physical walls.

Data centers host infrastructure, but the threats they face originate across software stacks, user identities, third-party integrations, and automated systems. Attackers no longer need physical proximity. They require only access, patience, and opportunity.

Recent large-scale exposures of billions of login credentials illustrate this reality. These breaches were not caused by a single failure. They emerged from weak authentication practices, compromised external platforms, credential reuse, and delayed detection. The aggregation of data within centralized systems magnified their impact.

Similarly, breaches involving identity and access management platforms, including single sign-on systems, revealed how attacks on authentication layers can cascade rapidly. When identity becomes the gateway to services, its compromise undermines every downstream control.

For data centers, this creates a shared responsibility problem. Infrastructure may be secure, but hosted platforms introduce vulnerabilities. Operators must balance isolation with flexibility, security with performance, and monitoring with privacy.

To manage this, modern facilities adopt layered defence models. Network segmentation limits lateral movement. Zero-trust principles remove implicit trust within internal networks. Encryption ensures that even accessed data remains protected. Behavioural analytics monitors anomalies rather than relying solely on signature-based detection.

Yet technology alone is insufficient. Many breaches still originate from human behaviour. Phishing attacks exploit trust. Misconfigurations expose services. Delayed patching leaves known vulnerabilities open. Cybersecurity remains, at its core, a human challenge expressed through digital means.

Physical Security: Risk in Plain Sight

As much as cyber threats take the center stage in strategic discourses, physical security still forms the foundation of all other security measures. A data center is a physical property. Its servers, power systems, and network equipment exist in space. Physical access remains one of the most direct and dangerous attack vectors.

Modern facilities are designed as layered environments. Perimeter fencing establishes the first boundary. Vehicle barriers reduce forced entry. The surveillance systems run around the clock, and they are backed by analytics that detect abnormal movement patterns. Access control is a combination of biometric verification and role-based permissions. It ensures that a person can enter only the authorized areas at authorized times.

Sensitive zones within the facility add further layers. Mantrap systems prevent tailgating. Power rooms, network cores, and control centers must be separately authorized. Logs capture all the entries and all the exits, thus giving audit trails in aid of accountability.

But physical security risk is not associated only with intrusion. Threats to the environment are also important. Operations can be destabilized by fire, flooding, seismic activity, as well as extreme weather. The high frequency of heatwaves, storms, and flooding in areas that were previously seen as stable has been exacerbated by climate change.

The response of facilities is in the form of elevated construction, superior drainage, seismic reinforcement, and fire-suppressing mechanisms that do not result in collateral damage to equipment. These are very expensive measures, and a lack of them would be disastrous.

A physically weak data center cannot be assured of digital safety. The two are inseparable. Standards like ISO/IEC 27001 & 27002, NIST SP 800-53/800-171, ANSI/TIA-942, EN 50600 (EU), PCI DSS define several physical security controls.

Supply Chains and the Hidden Dependencies

Reliability assumes the availability of components. This has been a very weak assumption.

Servers, networking devices, power electronics, cooling systems, and control programs are dependent on world-spanning global supply chains. When one area is interfered with, it has a ripple effect and slows the construction, upgrades, and repairs of another area.

The weakness of this model was revealed through the shortages of semiconductors experienced in the past years. Turnaround time on important parts increased drastically.

Spare parts, which used to take a few days to reach the company, were no longer available for months. There was a need to run equipment life cycles longer, which increased the risk of failure and maintenance load on operators.

Supply chains are also complicated by geopolitical tensions. Access to certain technologies can be limited by export controls, sanctions, and trade disputes. The reliance on a small group of manufacturers leads to a concentration risk, especially for high-performance processors and specialized networking equipment.

In response to this, operators are diversifying suppliers, standardizing hardware platforms, and holding strategic inventories of critical spares. Hyperscalers, in some cases, work on some custom hardware to become less dependent on vendors. Other companies decentralize some of their supply chain to reduce the risks of cross-border interference.

Society has embraced supply chain resilience as a fundamental aspect of reliability planning and not a logistical add-on.

Geopolitical and Regulatory Risk

There is a political environment that surrounds data centers, which determines the risk profile of a data center. The laws on data sovereignty, localization, trade regulations, and even national security issues affect the location of the facilities and the manner in which they are run.

Geopolitical unpredictability may interfere with the power supply, cross-border connectivity, as well as procurement of equipment. The changes in regulations have the power to transform the cost structures within a few hours. Policy modifications have impeded or stopped the development of data centers in some areas, either because of energy consumption or land utilization.

This presents uncertainty in the long run to operators. Data centers are constructed over a span of decades. The political priorities usually fluctuate over the years.

This risk needs to be managed with involvement as opposed to evasion. The operators collaborate with governments, utilities, and regulators to ensure that infrastructure development is in line with the national goals. The trust is developed through long-term power contracts, renewable energy relations, and openness of the environmental impact.

The risk of politics cannot be done away with, but it can be predicted and planned.

Cost Pressure and the Reliability Trade-off

Data centers are capital-intensive in nature. The investment in land, construction, power infrastructure, cooling systems, as well as network connectivity is expensive at the initial phase. The returns are achieved gradually and in most cases over a period of time.

This poses a strain on the control of costs versus reliability. Caution should be exercised to avoid redundancy or maintenance, or cutting staffing levels. Excessive investment without a tight rein on demand can put financial survival into jeopardy.

Another challenge is on energy costs. Electricity is one of the highest operational costs. Financial models are influenced by price volatility, carbon pricing, and sustainability requirements. Switching to renewable energy may have to be accompanied by extra capital investment, although it is set to offer stability in the long term.

Operators who succeed in it perceive cost management as a strategic activity. Efficiency investments lower the long-run operating costs. Gradual growth matches the demand. Contract structures share risk between operators and customers.

Financial discipline, in this context, becomes a reliability measure.

Human Capital Shortages

The scarcity of skilled human capital is probably the most underrated threat to the industry.

The contemporary data centers demand experts with knowledge of electrical networks, mechanical infrastructures, computer networks, cybersecurity, automation, and safety measures. These are specialized skills that are not easy to transfer. The increase in global expansion has exceeded the development of the workforce. As per Uptime Institute survey 2022, 54% of data centers had difficulty in finding qualified staff.

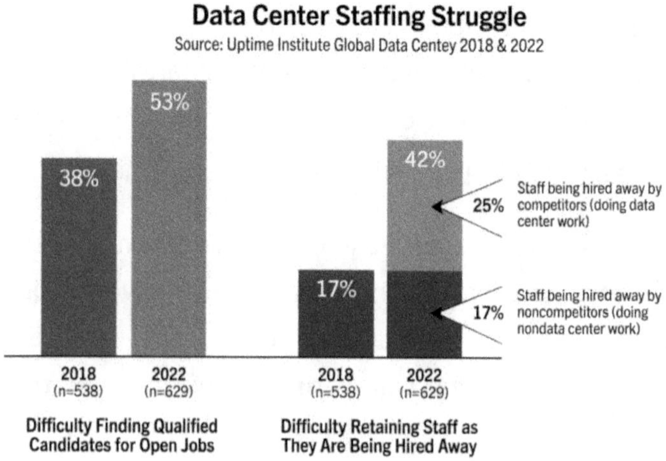

Figure 5.3 Staffing Struggle

The outcome is that talent has become a major winner-takes-all competition, labor expenses are climbing, and more and more companies are depending on rapid-track training. However, experience cannot be summed up. The process of judgment is acquired, and in most cases, through failure.

Operators counter it through investment in culture, retention, and training. Apprenticeship models enable the new entrants to train under supervision. Skills are kept up to date with continuous certification. Career progression is also minimised through clear career paths, and the memory of institutions is retained.

Automation minimizes human intervention; however, it does not eliminate human judgment. Capability is magnified by technology. It is not a replacement for responsibility.

Interdependence and Cascading Risk

The threat posed by these risks is that they have a tendency to interact. Security upgrades can be delayed because of a supply chain delay. Staffing is constrained by cost pressure. The shortage of skills enhances the chances of error during configuration. Cyber-attacks cause regulatory investigations, which influence financial soundness.

Data center risk is systematic as opposed to isolated.

It is necessary to understand this interdependence. The reliability does not come up because of the individual risks being removed, but through the effectiveness of their interactions.

Failure as Instruction

There is a tendency, particularly in highly engineered industries, to treat failure as an anomaly. Something to be explained away, documented briefly, and moved past as quickly as possible. Data centers, however, do not have the luxury of forgetting their failures. At scale, even minor disruptions leave traces that resurface later, sometimes in unexpected forms.

In this industry, failure is not merely endured. It is studied.

The most resilient data center ecosystems in the world are those that have learned how to extract structure from disruption. Their systems, processes, and cultures carry the imprint of past breakdowns, quietly shaping decisions long after headlines fade.

Japan's Earthquake-Resilient Data Centers: A Case Study for Reliable Design

Few environments test reliability more relentlessly than Japan. Positioned along the Pacific Ring of Fire, the country experiences thousands of seismic events every year. Earthquakes are not outliers. They are part of daily reality.

For data center operators in Japan, the question is not whether an earthquake will occur, but how often, how strong, and how prepared the facility will be when it does.

This awareness has shaped a distinct approach to resilience. Japanese data centers are not merely reinforced. They are designed to move.

Base-isolation systems separate the building structure from ground motion, allowing the earth to shift beneath without transmitting full force upward. Dampers absorb energy. Structural frames are engineered to flex rather than fracture. Inside the facility, server racks are anchored with seismic bracing, cable trays are designed with slack to accommodate movement, and cooling pipes incorporate flexible joints to prevent rupture.

Power systems receive similar attention. Fuel tanks are stabilized to prevent sloshing. Backup generators are mounted with vibration controls. Electrical panels are secured against lateral movement. Redundancy is not theoretical. It is tested repeatedly.

Equally important is operational readiness. Earthquake response drills are conducted routinely. Staff are trained not only to evacuate safely, but to return quickly and assess system integrity. Checklists guide post-event inspections. Control rooms prioritize verification of power quality, cooling stability, and network connectivity before resuming normal operations.

These practices have yielded measurable results. During major seismic events, including the 2011 Tōhoku earthquake, multiple data centers continued operating despite widespread damage to surrounding infrastructure. Services remained available not because earthquakes were avoided, but because disruption was anticipated and absorbed.

This approach reflects a deeper philosophy, articulated succinctly by engineering historian Henry Petroski, who observed, "The best engineers are not those who prevent failure, but those who design systems that fail well." In Japan's data centers, failure is not denied. It is managed with precision.

Global Cloud Outages: When Scale Magnifies Fragility Page

If earthquakes test physical resilience, large-scale cloud outages test systemic assumptions. In recent years, incidents affecting global platforms operated by companies such as AWS and Meta have demonstrated how complexity can undermine even the most advanced infrastructures.

These outages were not the result of single-component failures. They emerged from interactions within highly automated systems. A routine configuration change propagated across regions faster than anticipated. Dependency chains created feedback loops. Automated recovery systems responded as designed, yet their speed amplified the disruption by acting on incomplete information.

In one widely examined incident, internal network changes triggered congestion that cascaded across availability zones. In another, a software update affected internal tools used to diagnose and resolve the outage, slowing recovery. The irony was stark. Systems designed to ensure resilience became obstacles when conditions deviated from expected patterns.

What followed these incidents was as important as the failures themselves. Operators published detailed post-incident analyses. Architectural changes introduced stronger isolation between regions. Change management processes were tightened. Canary deployments and staged rollouts became more conservative. Automated systems were redesigned to allow for greater human oversight during anomalies.

These events reshaped industry thinking. Reliability could no longer be assessed solely at the facility level. It had to be evaluated across distributed software systems, organizational processes, and human decision-making.

Failure, in this context, became a catalyst for humility.

Training Models in Practice: Singapore and India

Different regions have responded to the human capital challenge in different ways, shaped by their constraints.

Singapore, limited by land and power availability, has focused on operational excellence. Data center roles are treated as specialized professions. Industry needs are matched with certification pathways. Constant training makes one aware of the changing technologies. Operators also work hand in hand with the government agencies in order to ensure their workforce is highly skilled.

The training focuses on accuracy and responsibility. Technicians are provided with training on how to read data from advanced monitoring systems and how to act proactively. The progression of careers is organized, and the attrition is minimized, and knowledge is retained in the institution.

India, by contrast, faces a scale challenge. Rapid expansion has created demand for thousands of professionals across multiple cities. The response has been diversification and partnership.

There has been an emergence of public-private projects to counter the shortage of skills. The areas of training programs include basic skills, electricity, cooling, safety measures, and working discipline. The models of apprenticeship enable new entrants to study under supervision and slowly absorb responsibility.

The methods vary, although the concept is the same. Reliability depends on people who understand not only systems, but also consequences.

From Incident to Institutional Memory

What distinguishes mature operators is not the absence of failure, but the presence of learning loops.

After every incident, detailed reviews are conducted. Timelines are reconstructed. Decisions are examined. Assumptions are questioned. These reviews are not exercises in blame, but in understanding.

Findings are translated into design changes, process updates, and training revisions. Over time, this creates institutional memory. Lessons persist even as individuals move on.

In organizations that value reliability, failure is not buried. It is integrated.

The case studies explored here illustrate a central truth. Resilience is not achieved by resisting disruption, but by engaging with it intelligently.

Earthquakes cannot be prevented. Complex systems will fail in unexpected ways. Cyber threats will evolve. Skills shortages will persist. What changes is the capacity to respond, adapt, and improve.

Failure has taught the industry much. Continuity is how those lessons are carried forward.

Disaster Recovery Beyond the Comfort of Backups

Disaster recovery is often spoken about in reassuring terms. Data is backed up. Systems are mirrored. Failover sites exist. The language suggests certainty, as though recovery is a mechanical reversal of failure.

Reality is less tidy.

Backups are essential, but they are only the beginning. A backup that cannot be restored within a meaningful timeframe is an archive, not a recovery mechanism. A failover system that has never been tested under load is an assumption, not a safeguard.

Modern data centers are designed for partial loss as a normal condition. Individual components are expected to fail without escalating into system-wide disruption. Applications are architected to tolerate degraded performance rather than collapse. Data replication occurs continuously across zones and regions, reducing dependency on single locations.

In practice, this means that disaster recovery begins not with restoration, but with continuity. Traffic is rerouted. Workloads are redistributed. Users remain connected, often without awareness that anything has occurred.

Human intervention, when required, is deliberate rather than frantic. Operators validate system behaviour, confirm integrity, and step in only when automation encounters ambiguity. The absence of panic is itself a design outcome.

Business Continuity as Organisational Memory

Business continuity extends far beyond technical resilience. It encompasses authority, communication, and institutional knowledge.

During major incidents, confusion can be as damaging as failure. Unclear escalation paths delay response. Conflicting decisions compound disruption. Silence breeds speculation, internally and externally.

High-reliability data center organisations address this through clarity. Roles are defined in advance. Decision thresholds are explicit. Incident commanders are appointed. Communication protocols specify who speaks, when, and to whom.

These structures are rehearsed regularly. Drills simulate not only technical failure, but organisational stress. Teams practice coordination across functions, operations, security, legal, communications, and executive leadership. The goal is familiarity, not heroism.

Fatigue is treated as a risk factor. Long-duration incidents strain attention and judgment. Shift rotations, redundancy in leadership roles, and mental health considerations are built into response planning. Continuity is sustained by people who are able to think clearly under pressure, not by those who endure exhaustion.

Learning Loops That Do Not Close

One of the defining traits of resilient data center operations is the refusal to treat incidents as closed chapters.

After-action reviews are conducted with rigor. Logs are examined in detail. Timelines are reconstructed minute by minute. Decisions are evaluated in context, not hindsight. Where procedures fail, they are revised. Where assumptions proved false, they are challenged openly.

Crucially, these reviews are shared. Lessons are documented and circulated across teams and facilities. Training programs are updated. Design standards evolve. Institutional memory is reinforced.

This process prevents repetition not by imposing rules, but by building understanding. Over time, organisations develop an intuitive sense of how systems behave under stress. Patterns are recognized earlier. Responses become more measured.

Reliability improves not through perfection, but through the accumulation of insight.

Human Capital as the Final Redundancy Layer

Across every discussion of continuity, one truth persists. When systems encounter conditions, they were not designed to handle, people compensate.

It is the technician who notices an anomaly that dashboards overlook. The engineer remembers a similar incident year earlier. The operations manager who senses escalation before alarms trigger.

This is why human capital cannot be treated as a variable cost to be optimized away. It is the final redundancy layer, the one that remains when automation reaches its limits.

Investment in people rarely produces immediate metrics. Its returns are visible only when something goes wrong, and nothing worse follows. Faster recovery. Quieter incidents. Preserved trust.

This reality echoes the observation of safety researcher Sidney Dekker, who wrote that "Technology changes quickly, but human judgment remains the last line of defence." In data centers, this line is crossed daily, often without recognition.

Training, retention, and culture are therefore not peripheral concerns. They are central to reliability.

Trust as the Invisible Outcome

At the end of this chapter, reliability reveals itself not as a technical achievement, but as a form of trust.

Users trust that systems will be there when needed. Businesses trust that operations will continue without interruption. Governments trust that critical services will remain available under stress.

This trust is fragile. It is built slowly and lost quickly. It depends on transparency as much as performance. Organisations that acknowledge failure, explain causes, and demonstrate learning preserve credibility even in disruption.

Those that obscure or deflect invite suspicion.

Data centers operate in this delicate space. Their success is measured not by attention, but by its absence. Years of flawless operation may pass unnoticed. A single incident may dominate perception.

What sustains trust over time is not perfection, but preparedness.

The Quiet Discipline That Endures

Reliability at scale is not accidental. It is the outcome of countless decisions made when nothing appears urgent. Choosing redundancy over efficiency. Training over shortcuts. Documentation over improvisation. Reflection on denial. These choices accumulate. They form an operating philosophy that persists even as technology changes.

As the book moves forward into discussions of artificial intelligence, edge computing, and next-generation architectures, the lessons of this chapter remain relevant.

Complexity will increase. Scale will expand. Automation will deepen.

Yet the fundamentals will not change.

Data centers, for all their machines and algorithms, endure because of discipline, judgment, and people who understand that continuity is not a response to crisis, but a way of working.

In that quiet discipline lies the true resilience of the digital world.

Chapter 6

The Future of Data Centers

"The future is not something we enter. The future is something we create." - Leonard I. Sweet.

The future of data centers is regarded as a place that one is expected to get to. A future where technology is faster, buildings are greener, and the systems are smarter. This mode of thinking is reassuring; however, it is deceptive.

We are not moving towards the future of data centers. It is a process that is already in motion.

Each design choice made today, the location of a facility, its power supply, workload control, and data control, determines the lines of what will be known ten years down the road. There will be no special day of arrival. Instead, there will be accumulation. Layer upon layer of choices, constraints, innovations, and compromises.

They are no longer peripheral infrastructure. They are not simply buildings that support the digital economy. They are the digital economy's physical expression. The concentration of computation, the distribution of intelligence, the management of energy, and the governance of data all converge within them. This convergence is what makes their future consequential.

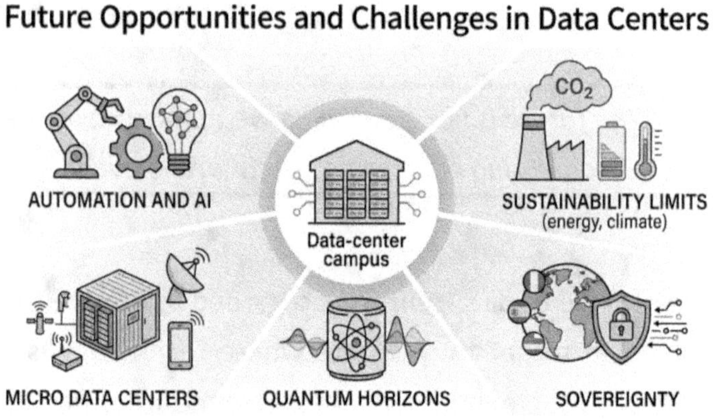

Figure 6.1 Future Challenges and Opportunities

From Static Facilities to Adaptive Systems

For much of their history, data centers were engineered for predictability. Load patterns were relatively stable. Enterprise-driven workloads were prevalent. Growth was slow but consistent. It was possible to plan infrastructure several years ahead with reasonable certainty.

That stability has disappeared.

Modern workloads are volatile. Artificial intelligence education may require colossal computational surges. Live stream channels have bursts that are associated with world events. Finance systems have round-the-clock operations operating even across time zones. Internet-of-Things devices produce data continuously, and in many cases, it is random.

Static design no longer suffices.

The future data center must be adaptive. Not only resilient in the meaning of being able to withstand failure, but also responsive in the meaning of self-balancing in a continuous manner. The rerouting of power should be dynamic. Cooling must adjust in real time. Network paths must evolve as traffic shifts. Compute resources must be allocated based on changing priorities.

This is not a cosmetic evolution. It represents a philosophical shift in infrastructure design.

The data center is no longer a fixed environment hosting variable activity. It is a variable environment responding to variable demand.

Artificial Intelligence as an Operational Layer

At the center of this transformation lies artificial intelligence, not as a workload, but as an operating layer.

Early uses of AI in data centers focused on narrow optimisation tasks. Cooling efficiency. Power management. Predictive maintenance. These applications demonstrated value, but they were isolated.

The future lies in integration.

It is becoming widespread to use AI systems to organize complete facilities. They consume telemetry of thousands of sensors, historic performance, workload projections, and energy price indications. They make decisions continuously, often without direct human instruction.

In effect, the data center begins to manage itself.

This autonomy is not absolute. Humans define constraints, priorities, and ethical boundaries. But within those boundaries, systems act with a level of speed and complexity no operations team could match manually.

This capability fundamentally alters risk management. Instead of preparing only for known failure modes, operators can model probabilities, anticipate cascading effects, and intervene earlier. Reliability becomes proactive rather than defensive.

The implications extend beyond efficiency. They change staffing models, training requirements, and governance structures. The human role shifts from operator to overseer, from responder to designer.

Intelligence Inside the Building

Artificial intelligence does not function in abstraction. It depends on data, and data depends on perception.

Future data centers are dense sensory environments. Temperature, humidity, vibration, airflow, power quality, and equipment health are monitored at granular levels. This sensory fabric creates situational awareness.

Small deviations matter.

A barely perceptible change in airflow can indicate obstruction. A subtle shift in vibration frequency can signal mechanical fatigue. A marginal voltage fluctuation can precede component failure. AI systems trained on years of operational data learn these patterns.

What emerges is not just monitoring, but memory.

The facility remembers its own behaviour. It learns from past stress, adapts to new conditions, and refines its responses over time. This capacity for learning distinguishes future data centers from their predecessors.

They do not merely operate. They evolve.

The Collapse of Latency Tolerance

While intelligence transforms how data centers function internally, external forces are reshaping where and why they are built.

Latency tolerance is collapsing.

The delay cannot be afforded in applications that characterize the upcoming decade of digital services. Immediate response is required with autonomous systems, immersive media, real-time analytics, remote medical procedures, and industrial automation.

The distance is used as a weakness.

No matter how powerful, traditional centralized architectures are, they cannot fulfil these requirements on their own. The processing of data has to be brought as close as possible to the generation and consumption points. This necessity gives rise to edge computing as a structural component of future infrastructure.

Edge data centers are not replacements for hyperscale facilities. They are complements. They constitute a layer architecture in which various classes of work are processed at varying degrees of proximity to the user.

The future is neither centralized nor decentralized. It is coordinated.

A Layered Infrastructure Model

In this emerging model, hyperscale data centers continue to anchor global platforms. They train large AI models, store massive datasets, and provide economic gravity. Edge facilities handle time-sensitive processing, regional services, and local decision-making.

Between these layers flows data, constantly evaluated and routed based on urgency, cost, energy availability, and risk.

This coordination is itself managed by intelligence.

Workloads shift dynamically. Non-critical processing moves to locations with surplus renewable energy. Latency-sensitive tasks stay local. Redundancy is distributed rather than duplicated.

Infrastructure becomes fluid.

This fluidity improves resilience. Failure in one node does not cascade catastrophically. Systems reroute. Services degrade gracefully rather than collapsing.

The architecture resembles a network more than a hierarchy.

Rethinking Scale and Growth

One of the most persistent misconceptions about the future of data centers is that growth simply means building bigger facilities.

Scale still matters, but it is no longer the sole metric.

Repeatability, flexibility, and integration matter as much as raw capacity. Modular construction, standardized designs, and prefabricated components allow capacity to be deployed incrementally. Growth aligns more closely with demand.

This approach reduces speculative risk. It also enables faster adaptation to regulatory changes, energy constraints, and technological shifts.

In this sense, the future data center is less a monument and more an organism.

What emerges from these trends is a clear inflection point.

Data centers are crossing a threshold from being engineered environments to being adaptive systems. From static infrastructure to intelligent networks. From distant utilities to embedded societal assets.

This transition is still incomplete. Many facilities operate under older assumptions. Many policies reflect past realities. But the direction is unmistakable.

Data centers have outgrown their original design assumptions.

What follows is an exploration of how this transformation unfolds across geography, energy systems, governance, investment, and society itself. The future of data centers is not just technical. It is economic, political, and human.

Edge, Decentralization, and the End of Distance

The most visible shift in the future of data centers is not happening inside the buildings. It is happening around them.

For nearly two decades, the dominant logic of digital infrastructure was concentration. Compute the power gathered in large campuses. Storage aggregated in fewer locations. Connectivity radiated outward from a handful of hubs. This model delivered scale, cost efficiency, and operational simplicity. It was the right answer for its time.

But the conditions that made centralization optimal are changing.

Data is no longer generated primarily by users sitting at desks. It flows from machines, vehicles, cameras, medical devices, sensors, and autonomous systems. It is continuous, local, and time-sensitive. Sending all of it to distant hyperscale facilities is neither efficient nor, in many cases, possible.

The future, therefore, demands a redistribution of computation, not away from large data centers, but alongside them.

Why Distance Became the Bottleneck

Latency was once an engineering detail. A few hundred milliseconds of delay rarely altered user experience in early web applications. Today, milliseconds define functionality.

Autonomous vehicles must respond faster than human reflexes. Industrial automation depends on real-time feedback loops. Online gaming, augmented reality, financial trading, and telemedicine all operate at speeds where delay becomes failure.

Bandwidth alone cannot solve this problem. Even the fastest networks are bound by physics. Distance introduces delay, and delay introduces risk.

This is where the edge enters, not as a convenience, but as a necessity.

Edge data centers bring computation closer to the point of data creation. They reduce round-trip time. They allow local decisions to be made locally. They act as buffers, filters, and control points within larger systems. The future of data centers must therefore be understood as a system of layers, not a hierarchy of size.

Defining the Edge Without Romanticism

Edge computing is often described loosely, sometimes inaccurately. It is not simply about small data centers. Nor is it about abandoning the cloud.

The edge is a functional position, not a physical one.

An edge data center is any facility where computation occurs closer to the source of data than traditional central infrastructure. This may be a small modular unit near a telecom exchange, a facility embedded in an industrial campus, or a regional data center serving a cluster of cities.

They handle tasks that cannot tolerate delay or disconnection. Everything else can still flow to larger centers. The edge data center market is projected to grow from about USD 16.9 billion in 2026 to roughly USD 71.9 billion by 2035.

In this sense, the edge does not compete with hyperscale infrastructure. It completes it.

Micro Data Centers as Infrastructure Units

One of the defining features of edge infrastructure is modularity.

Future data centers will increasingly be built as repeatable units rather than bespoke projects. Micro data centers, often containerized or prefabricated, can be deployed rapidly. They require less land, less time, and fewer approvals. They scale horizontally rather than vertically.

This approach changes how infrastructure grows.

Instead of planning for peak demand years in advance, capacity can be added incrementally. Instead of concentrating risk in a single location, it is distributed across nodes. Failure becomes local rather than systemic.

These facilities are designed to operate with minimal human presence. Remote monitoring, AI-driven diagnostics, and standardized components reduce operational complexity. Maintenance becomes modular. Expansion becomes predictable.

This logic aligns with how digital platforms scale. Infrastructure begins to mirror software in its deployment philosophy.

Telecommunications and the Edge Convergence

The rise of edge data centers is inseparable from the evolution of telecommunications.

5G networks do more than increase speed. They enable network slicing, ultra-reliable low-latency communication, and massive device density. These capabilities only deliver value when paired with nearby compute.

As a result, telecom operators are becoming infrastructure partners in new ways. Data centers are being integrated into network architecture. Compute resources sit alongside switching equipment. Network and compute planning converge.

This convergence blurs traditional boundaries.

Telecom facilities become computational nodes. Data centers become network elements. The distinction between connectivity and computation fades.

This integration also changes business models. Revenue shifts from pure bandwidth to value-added services. Operators participate directly in digital ecosystems rather than merely transporting data.

The future data center is therefore as much a network component as a compute facility.

Industrial Systems and Local Intelligence

The best example of edge infrastructure, perhaps, is in the industrial setting.

The manufacturing plants, logistics centers, power plants, and transportation systems produce large amounts of data. The sensors check the state of equipment, the quality of production, the energy use, and the safety situation constantly.

It is not efficient and risky to send all this data to central facilities. There is a need to make decisions within milliseconds. Network disruptions cannot halt operations.

Edge data centers embedded within or near industrial sites process data locally. They detect anomalies, trigger interventions, and optimize operations in real time. Only aggregated insights or long-term records move upstream.

This model increases efficiency, safety, and resilience.

It also changes the relationship between digital infrastructure and physical production. Data centers move from supporting back-office systems to participating directly in operational control.

"The factory of the future runs on algorithms as much as machines." – Klaus Schwab.

In this environment, the edge is not optional. It is foundational.

Urban Systems and Real-Time Governance

Cities are becoming computational entities.

Traffic lights respond to congestion. Surveillance systems analyse patterns. Utilities balance load dynamically. Emergency services rely on live data. All of this requires immediate processing.

Centralized systems struggle under this demand. Latency introduces risk. Network congestion degrades service.

Edge data centers provide local processing capacity for urban systems. They enable real-time decision-making without constant reliance on distant infrastructure. They support continuity during network disruptions.

This capability becomes increasingly important as cities adopt smart infrastructure. Governance shifts from periodic planning to continuous management.

The data center becomes part of the city's nervous system.

The Economic Logic of Decentralization

Decentralization is often framed as a technical necessity. It is also an economic one.

Building massive hyperscale campuses requires significant capital, long lead times, and regulatory clarity. Edge facilities can be deployed faster, closer to demand, and with lower upfront investment.

They open infrastructure participation to more regions. Smaller cities attract digital activity. Local enterprises benefit from improved performance. Economic value spreads geographically.

This redistribution aligns with broader economic goals in many countries, including regional development and digital inclusion.

Infrastructure ceases to be a winner-takes-all asset.

India's Tier-2 and Tier-3 Shift and the Rewriting of Digital Geography

For much of India's digital ascent, progress followed a familiar pattern. Innovation clustered in a few metropolitan centers. Infrastructure flowed toward capital, talent, and connectivity that already existed. Mumbai, Chennai, Hyderabad, and Delhi-NCR became natural anchors for data centers because they offered undersea cables, stable power corridors, mature real estate markets, and regulatory familiarity.

This concentration delivered speed. It also created an imbalance.

As digital services became essential rather than optional, the limits of metro-centric infrastructure began to show. Latency increased with distance. Network congestion affected quality. Smaller cities and rural regions remained dependent on far-off compute capacity for services that now touched daily life, payments, healthcare, education, and governance.

The future of data centers in India is therefore inseparable from a geographic shift.

Edge and decentralized infrastructure are not simply technological upgrades. They represent a reordering of where digital value is created and who participates in that creation.

Emerging Tier-2 Nodes and Their Roles

Several Indian cities have begun to assume distinct roles within this decentralized landscape.

Pune functions as an industrial and automotive edge, supporting manufacturing analytics, enterprise SaaS platforms, and proximity workloads that complement Mumbai's hyperscale capacity. Kochi leverages submarine cable connectivity to serve regional cloud traffic and disaster recovery functions. Ahmedabad supports fintech platforms, state-level digital services, and logistics systems tied to western India's trade corridors.

Cities such as Jaipur, Indore, Coimbatore, Nagpur, and Bhubaneswar increasingly host smaller facilities designed to serve surrounding districts rather than global markets. These are not prestige projects. They are utility deployments.

What distinguishes these centers is not scale, but relevance.

They process Aadhaar authentication requests, UPI transactions, healthcare records, educational content, and municipal data closer to where it is generated. They reduce dependency on long-haul networks for everyday operations.

This ordinariness is precisely what makes the shift transformative.

Local Infrastructure, Local Confidence

One of the less visible effects of decentralized data centers is psychological.

When infrastructure is distant, digital systems feel abstract. When it is nearby, they feel tangible. This tangibility breeds confidence.

Local enterprises are more willing to digitise operations when latency is predictable and support is accessible. Start-ups build region-specific solutions when performance constraints are manageable. Educational institutions adopt hybrid models more readily. Hospitals integrate diagnostics and telemedicine with fewer reliability concerns.

In this way, data centers function as a confidence infrastructure.

They lower perceived risk. They reduce friction. They make digital ambition feel attainable beyond major cities.

This effect compounds over time. Each successful deployment encourages the next.

State Policy as a Catalyst

India's federal structure plays a critical role in shaping this future.

States increasingly recognise data centers as enabling infrastructure rather than niche real estate assets. Policy frameworks have begun to evolve accordingly.

Tamil Nadu, Telangana, Maharashtra, Uttar Pradesh, and Gujarat have introduced data center policies that explicitly encourage non-metro deployment. Incentives extend beyond capital regions. Power tariffs, land use classifications, and approval processes are being adapted to support distributed builds.

This policy shift matters.

Edge infrastructure depends on predictability more than subsidy. Clear zoning, reliable power access, and streamlined permissions reduce deployment risk. When states provide these conditions, investment follows.

The future map of India's data centers will therefore reflect policy clarity as much as demand density.

Municipal Systems and Proximity Governance

As computing power moves closer to towns and districts, governance models adapt.

Municipal systems increasingly rely on real-time data. Traffic management, water distribution, waste collection, emergency response, and public safety platforms depend on continuous processing. Edge data centers allow these systems to function with greater autonomy and resilience.

Local decision-making improves. Dependence on central systems reduces. Service continuity strengthens during network disruptions.

This decentralisation of computation mirrors decentralisation of governance.

Decisions are made closer to the conditions they address. Data stays nearer to its source. Accountability becomes more direct.

"The closer a decision is to its consequences, the better it tends to be." – Elinor Ostrom.

In this sense, edge infrastructure supports democratic capacity as much as technical efficiency.

Economic Spillovers Beyond Construction

The economic impact of data centers in Tier-2 cities extends beyond construction and operations.

Training ecosystems emerge. Technical schools are changing their curriculum to incorporate the operations of the data centers, power systems, cooling systems, and network management. The service providers specialise in maintenance, security, and logistics. Local working positions stabilize.

More to the point, digital entrepreneurship becomes diversified.

Start-ups do not feel the urge to move to metros just to have access to infrastructure. Regional platforms are formed, depending on local languages, markets, and needs. There is less homogenisation in innovation.

This diversification empowers the national digital economy by diminishing the systemic reliance on several hubs.

Risks and Coordination Challenges

Decentralization introduces complexity.

Distributed facilities increase the attack surface for cybersecurity threats. Physical security standards vary by location. Talent availability remains uneven. Operational coordination becomes more demanding.

These risks do not negate the model, but they demand standards.

National frameworks for security, interoperability, and monitoring become essential. Decentralization without coordination would fragment resilience rather than enhance it.

The future, therefore, lies in managed distribution.

Common protocols, shared best practices, and central oversight ensure that local autonomy strengthens the system as a whole.

A New Mental Map of Infrastructure

Perhaps the most profound shift underway is conceptual.

India's data center future will not be defined by a single megaproject or policy announcement. It will be shaped by hundreds of smaller decisions across states, cities, and districts.

Edge infrastructure marks a turning point because it aligns national digital ambition with regional reality.

It allows growth to spread rather than concentrate. It enables participation beyond metros. It grounds the digital economy in physical communities.

This shift does not diminish the importance of hyperscale facilities. It complements them. Together, they form a network that reflects India's scale, diversity, and complexity.

Hyperscale, Automation, and the Rise of the AI-Native Facility

Hyperscale data centers form the gravitational center of the edge infrastructure, in case it signifies the external dispersion of computation. They are the largest models that are being trained, the global platforms are coming to life, and the economic logic of scale is still indisputable.

Yet hyperscale data centers are not static monuments to bigness. They are undergoing a transformation as profound as that of edge facilities, driven by a single pressure: complexity has exceeded human manageability.

The future hyperscale facility is not defined only by size or power consumption. It is defined by autonomy.

Between the Monitoring and the Orchestration

Intelligence is not located on the edge in an AI-native hyperscale facility. It is entrenched in the fundamental procedures.

The telemetry power infrastructure, cooling systems, network fabrics, workload schedulers, and environmental sensors are ingested by AI systems. They are used to correlate signals across domains, learning the interactions of changes in one layer propagating to other layers.

Instead of alerting humans to problems, they reconfigure systems automatically. Workloads migrate. Power paths shift. Cooling patterns adjust. Failures are isolated before they cascade.

This orchestration operates continuously.

The data center ceases to be a collection of systems managed in parallel. It becomes a single, integrated organism.

This coordination changes the economics of scale. Reliability improves not because redundancy increases, but because intelligence reduces unnecessary stress on the system.

Self-Healing Infrastructure

One of the defining traits of future hyperscale facilities is self-healing capability.

In traditional environments, failure triggers alarms, investigations, and manual intervention. In AI-native facilities, failure triggers adaptation. Systems learn from the event, update predictive models, and adjust future behaviour.

A cooling unit failure may lead to temporary workload redistribution rather than emergency repair. A network congestion pattern may prompt proactive rerouting before performance degrades. A power anomaly may result in recalibrated load balancing across the campus.

The facility develops institutional memory.

Over time, this learning compounds. The data center becomes better at surviving its own stresses. Downtime decreases not because components stop failing, but because failure becomes expected and absorbed.

This shift reframes reliability.

Reliability is no longer the absence of failure. It is the presence of adaptation.

The Energy-Compute Feedback Loop

The most significant impact of AI-native hyperscale design is in energy management.

Facilities that are hyperscale use colossal amounts of energy. With pressure on grids over electrification, renewables integration, and climate variability, there is no longer a place to go with static energy planning.

The energy in the future data centers is not a constant input, but a dynamic variable.

AI systems forecast energy availability, pricing, and carbon intensity. Workloads are scheduled accordingly. Non-urgent processing shifts to periods of renewable surplus. Latency-tolerant tasks migrate across geographies based on grid conditions.

Computing power becomes mobile in time and space.

This feedback loop aligns economic efficiency with environmental responsibility. Energy-intensive AI training workloads move when power is cleanest or cheapest. Facilities reduce peak demand stress on grids.

In this model, hyperscale data centers act not as passive consumers, but as active participants in energy ecosystems.

Standardisation at Planetary Scale

Another defining feature of future hyperscale infrastructure is radical standardization.

To enable autonomy, components must be predictable. Hardware designs converge. Software stacks unify. Interfaces stabilize. Variation becomes the enemy of learning.

This standardization allows AI systems to generalize insights across campuses. A failure pattern observed in one geography informs preventive action elsewhere. Optimizations discovered in one facility propagate globally.

The hyperscale operator becomes a planetary infrastructure manager.

This scale of learning creates advantages that are difficult for smaller operators to replicate. It reinforces the dominance of a few global players, even as edge infrastructure decentralizes capability elsewhere.

The future, therefore, contains a paradox: decentralized compute at the edge, centralized intelligence at scale.

Human Roles in an Autonomous Environment

As hyperscale facilities become more autonomous, the role of human expertise changes.

Operators shift from executing procedures to defining constraints. Engineers focus on system design rather than incident response. Ethics teams evaluate decision boundaries rather than outputs.

The most critical human contribution becomes judgment.

What should the system optimize for? Cost, carbon, latency, resilience, or fairness? How should trade-offs be resolved when objectives conflict? Where must automation stop?

These are not technical questions alone. They are governance questions. The future hyperscale data center, therefore, elevates human responsibility rather than erasing it.

Risks of Over-Automation

AI-native operations are not without risk.

Opaque decision-making can obscure accountability. The error of models can be scalable. Excessive automation may undermine human intuition. The threats posed by cyber criminals change together with defensive intelligence.

Such dangers require management.

Auditability, transparent and fail-safe design, have become non-negotiable. The autonomous systems need to be explainable. Manual intervention paths must remain viable. Training must emphasize understanding, not blind trust.

The future is not fully autonomous infrastructure. It is supervised autonomy.

Hyperscale data centers are crossing a threshold where size alone no longer defines advantage. Intelligence does.

Facilities that cannot adapt will struggle under volatility. Those who learn will thrive.

This reshapes competition, investment, and regulation. It raises questions about the concentration of power, energy equity, and global dependency.

But it also unlocks unprecedented efficiency, resilience, and scale.

Quantum Horizons and the Quiet Rewriting of Computational Assumptions

Quantum computing is in an odd place in the discussion of data center futures. It is talked about both admiringly and suspiciously, as either something that is about to happen or something that is a far-off fantasy. Both views miss the point.

Quantum computing is not important because it will replace classical computation. It is important because it destabilizes the assumptions upon which today's infrastructure is built.

For data centers, the quantum era will not arrive suddenly. It will arrive asymmetrically, partially, and quietly, reshaping design choices long before quantum machines become widespread.

Why Quantum Matters to Data Centers Today

The current form of large-scale quantum computers will never be hosted in most of the existing data centers. Quantum systems require very special conditions, i.e., almost-zero temperatures, isolation of vibrations, electromagnetic isolation, and specialized power stability. These needs render them incompatible with traditional server halls.

Yet quantum computing already matters.

It matters because it challenges the foundations of cryptography. It matters because it introduces new workload types that must coexist with classical systems. It matters because it alters the strategic planning horizon of infrastructure operators.

The future data center must be quantum-aware long before it becomes quantum-capable.

This awareness influences architecture, security strategy, and investment planning.

Post-Quantum Security as an Infrastructure Imperative

The most immediate quantum impact lies in security.

Many encryption schemes that protect today's data are vulnerable, in principle, to sufficiently powerful quantum computers. While such machines do not yet exist at scale, the data they could compromise is already being generated and stored.

This creates a temporal vulnerability.

Efforts to encrypt sensitive data in the present day can be collected and decrypted in several years when quantum capabilities have advanced. This risk is not to be disregarded by governments, financial institutions, healthcare providers, and infrastructure operators.

Data centers, therefore, face a strategic shift toward post-quantum cryptography.

This transition is not trivial. Cryptographic algorithms are embedded deeply into systems, protocols, and hardware. Updating them requires coordination across software stacks, hardware vendors, and regulatory frameworks.

For data centers, this means designing systems that can evolve cryptographically without disruptive overhauls.

Security becomes a lifecycle concern rather than a static feature.

"The danger is not that quantum computers will break encryption tomorrow. The danger is that we are not preparing for when they can." – Whitfield Diffie.

Hybrid Compute Environments

Beyond security, quantum computing introduces the prospect of hybrid workloads.

Quantum machines excel at specific problem classes: optimization, simulation of quantum systems, and certain forms of pattern analysis. They are not general-purpose replacements for classical servers.

In the future, data centers may host quantum accelerators alongside GPUs and CPUs, much as GPUs were once specialized components that later became mainstream.

This hybrid model requires orchestration.

Workloads must be decomposed intelligently. Classical systems handle pre- and post-processing. Quantum systems tackle specific computational cores. Results recombine within broader workflows.

Data centers must therefore be designed for heterogeneity.

Power systems, cooling strategies, network fabrics, and scheduling algorithms must accommodate radically different compute profiles within a single environment.

This complexity reinforces the trend toward AI-native orchestration. Manual scheduling becomes impossible when workloads span classical, accelerated, and quantum domains.

Physical Design Implications

Even without hosting full-scale quantum systems, future data centers must account for their physical implications.

Some facilities will include dedicated quantum labs adjacent to classical infrastructure. Others will connect to external quantum services via secure, low-latency links. Still others will prepare space and power capacity for future retrofits.

These choices influence site selection, building layout, and long-term capital planning.

Operators must ask questions that would have seemed premature a decade ago. Where can vibration be minimized? How can extreme cooling be integrated without disrupting existing systems? How should electromagnetic interference be managed?

The answers vary, but the questions themselves signal a shift.

The future data center is planned with uncertainty in mind.

Quantum and the Redistribution of Advantage

Quantum computing also raises questions about geopolitical and economic concentration.

Quantum research is capital-intensive. It favors nations and corporations with long investment horizons and strong scientific ecosystems. This creates the risk of uneven capability distribution.

Data centers become strategic assets in this context.

Facilities that can integrate quantum capabilities, even partially, may offer services that others cannot. This amplifies existing disparities between global infrastructure leaders and followers.

For countries like India, the question is not whether to compete head-on in quantum hardware, but how to integrate quantum readiness into a broader digital infrastructure strategy.

This may involve partnerships, shared facilities, and policy support for hybrid research environments.

Data Centers as Long-Horizon Infrastructure

Quantum computing underscores a broader truth about data centers.

They are built for decades, not product cycles.

The modern construction of a hyperscale facility will probably maintain some form of operation within twenty years. The current choices regarding layout, power, cooling, and security will limit any future flexibility.

This long horizon distinguishes data centers from most technology assets. It also makes them uniquely sensitive to paradigm shifts.

Operators must therefore balance present efficiency with future optionality. Over-optimizing for today's workloads risks fragility tomorrow.

Managing Uncertainty Without Paralysis

Preparing for quantum does not mean waiting.

The mistake would be to defer decisions until uncertainty resolves. Infrastructure cannot be built on hesitation.

Instead, future-ready data centers adopt principles that accommodate unknowns:

- Modular design
- Cryptographic agility
- Heterogeneous compute support
- Flexible power and cooling systems
- Policy and governance frameworks that anticipate change

These principles serve quantum readiness, but they also serve resilience more broadly.

Quantum is not the only unknown ahead. New materials, new energy technologies, and new regulatory regimes will be introduced. One can adapt to an infrastructure that can adapt to others.

Quantum computing is a slight yet significant milestone in data centers.

It reminds us that infrastructure is not just about scaling known demands. It is about remaining viable under unknown futures.

Data centers that plan only for incremental improvement risk obsolescence. Those who plan for discontinuity gain strategic depth.

Quantum computing changes assumptions about security, computation, and time horizons.

It is a reality whether quantum machines will become ubiquitous or specialized, but their impact on data center design has already come to fruition.

Energy, Climate and the Hard Limits of Growth

"The real constraint on the future is not technology, but energy." – Vaclav Smil.

Every discussion about the future of data centers eventually arrives at the same immovable fact. Computation consumes power. And power, unlike software, is bounded by physics, geography, and climate.

Energy has long been considered as an input variable, the one that should be purchased relatively cheaply, reliably, and in adequate amounts. Sustainability was subsequently covered in terms of offsets, efficiency measurement, or compliance reporting. That sequencing is no longer viable.

Energy is no longer a background condition. It is the primary design constraint.

But energy is only the first of several hard limits. Cooling is now as decisive as compute density, turning thermal management into an architectural challenge rather than an operational one. Water availability, local climate, and heat rejection capacity increasingly determine where data centers can even exist. Sustainability, once framed as responsibility, has become a condition for permission, shaping regulation, financing, and public acceptance.

Beyond these, other physical boundaries assert themselves: land, grid stability, material supply chains, and the resilience of surrounding infrastructure.

The future of data centers will be shaped less by what technology can do and more by what energy systems and the ecosystems around them can sustain.

The End of Energy as an Afterthought

During the initial growth stage of cloud infrastructure, data center developers valued connectivity, availability of land, and tax incentives. Power followed. Grids were assumed to scale. Demand growth was incremental enough to absorb.

That assumption has collapsed.

The power consumption of AI is enormously high, especially for large-scale model training and inference. One hyperscale campus is able to use as much power as a small city. Take this and multiply it by dozens of facilities, and strain is bound to occur.

Utilities struggle to provision capacity quickly enough. Transmission infrastructure lags demand. Peak loads threaten grid stability. In some regions, new data center approvals are delayed or halted due to power constraints.

Energy has moved from a cost center to a bottleneck. The future data center must therefore be designed around energy realities, not aspirations.

Power Usage Effectiveness Is No Longer Enough

For years, Power Usage Effectiveness (PUE) served as the industry's primary efficiency metric. Lowering PUE became a badge of honour. Hyperscale operators drove it down through better cooling, airflow management, and equipment efficiency.

But PUE has limits.

It measures how efficiently power is used within a facility, not whether that power is sustainable, available, or socially acceptable. A perfectly efficient data center running on fossil fuels during peak grid stress still imposes real costs.

The future requires a wider perspective.

The quality of energy is as important as the quantity of energy. Carbon intensity, time-of-use alignment, water impact, and grid interaction are all applicable metrics.

Efficiency is necessary, but insufficient.

Synchronizing Compute with Energy

One of the most consequential shifts underway is the synchronization of compute workloads with energy availability.

Rather than treating power supply as static, future data centers treat it as variable. Renewable generation fluctuates with the weather and time. Energy prices change hourly. Grid stress varies seasonally.

AI-native orchestration systems already use these signals.

Non-urgent workloads are scheduled when renewable generation is high. Training jobs migrate geographically to regions with surplus clean energy. Latency-tolerant processing shifts to off-peak hours. Facilities throttle consumption dynamically to avoid peak demand.

Computing needs will become flexible. This flexibility transforms data centers from grid burdens into grid partners.

The future data center does not simply consume energy. It participates in energy markets.

Renewable Integration Beyond Offsets

Initial sustainability initiatives had high dependence on carbon offsets and renewable energy certificates. Although these mechanisms contributed, they did not do much in helping with real-time grid impact.

The future lies in direct integration.

On-site solar, wind partnerships, battery storage, and microgrids increasingly form part of data center design. Some facilities operate partially off-grid. Others coordinate closely with utilities to stabilize local networks.

Energy storage becomes critical. Batteries smooth out intermittency. They allow facilities to draw less from the grid during peak periods and more during surplus.

This integration requires capital and coordination. It also demands regulatory alignment. Energy markets must allow large consumers to act flexibly. Policies must support private generation and storage.

Where this alignment exists, sustainability accelerates. Where it does not, growth slows.

Water, Cooling, and Environmental Trade-offs

Energy is not the only environmental constraint.

Cooling systems consume water. In water-stressed regions, this creates conflict. Communities question whether digital infrastructure should compete with agriculture or drinking water.

The future data center must therefore rethink cooling.

Air cooling innovations, liquid cooling, immersion cooling, and heat reuse reduce water dependency. Some facilities recycle water extensively. Others relocate to climates where cooling demand is lower.

Heat itself becomes a resource.

Waste heat from data centers can warm buildings, greenhouses, or industrial processes. In colder regions, district heating systems already integrate data center output.

The future data center treats by-products as inputs elsewhere.

The Geography of Sustainable Growth

Energy limits are transforming the location of data centers.

Areas that have high renewable resources benefit. The areas that have high hydropower potential, plains with plenty of wind, deserts inhabited by the sun, and regions where cooling can be achieved by the sea become tempting.

On the other hand, there are limits to the regions where the grids are congested or when there is a shortage of water.

This geography has an impact on national strategies. Developments in energy ecosystems are not only a point of competition between countries based on policy incentives. The growth of data centers is in line with greater energy transformation objectives.

In this sense, the future of data centers cannot be separated from the future of national power grids.

Community Acceptance and Social License

Energy and environmental impact shape public perception.

Communities increasingly question large data center developments. They ask about power usage, water consumption, land impact, and local benefit. Projects that ignore these concerns face resistance.

The concept of social license becomes critical.

Future-oriented operators engage communities early. They invest in local infrastructure, renewable projects, and workforce development. They align data center benefits with local needs.

This engagement is not charity. It is risk management. Without trust, even efficient projects stall.

Climate Risk and Physical Resilience

Climate change introduces new risks.

Heatwaves increase cooling demand. Floods threaten facilities built near water. Wildfires disrupt power and connectivity. Extreme weather events test redundancy assumptions.

Future data centers must be climate-resilient.

Site selection accounts for long-term climate projections, not historical averages. Building design incorporates flood protection, heat tolerance, and supply chain resilience. Insurance models adapt.

Resilience planning becomes central rather than exceptional.

Energy and climate realities impose a hard truth.

There is no infinite digital growth without corresponding energy transformation.

The future of data centers will be characterized by the capability to become a part of the sustainable energy system, remain within environmental boundaries, and align with the expectations of society.

The ones that do will scale accountably. The ones that fail will be faced with opposition, control, and limitation.

Sovereignty, Policy, and the Architecture of Long-Term Choice

The future of data centers will not be decided inside server halls alone. It will be shaped just as forcefully in ministries, boardrooms, regulatory offices, and community meetings. Technology sets the direction, but policy and capital determine the pace, the form, and the limits.

Data centers sit at an unusual intersection. They are private assets that behave like public infrastructure. They depend on collective resources such as land, power, water, and spectrum. Their failures affect millions who may never know their names. This dual character makes governance unavoidable.

As the industry grows in scale and consequence, neutrality becomes an illusion. Every decision about data centers is a decision about national priorities.

Governments as Enablers and Gatekeepers

In the early years of data center growth, governments largely played a facilitative role. They offered tax incentives, eased land acquisition, and streamlined approvals. The goal was attraction. Data centers were seen as symbols of digital progress and foreign investment.

That phase is ending.

As energy demand rises, land scarcity intensifies, and environmental concerns sharpen, governments are moving from encouragers to stewards. The question is no longer how many data centers to attract, but which ones, where, and under what conditions.

Policy is becoming more selective.

Zoning regulations now account for grid capacity and water availability. Environmental clearances scrutinize lifecycle impact. Data localization laws influence where sensitive workloads reside. Energy policy determines whether growth is viable at all.

This selectivity is not an obstruction. It is maturity. The future of data centers depends on whether that playing field is coherent or fragmented.

National Strategy and Digital Sovereignty

Data centers have become instruments of national strategy.

Where data is stored, processed, and protected matters for economic competitiveness, security, and governance. Countries increasingly view domestic data center capacity as strategic infrastructure, comparable to ports or power plants.

This perspective reshapes policy priorities.

Data localization requirements influence facility placement. Cross-border data flow agreements affect global cloud architecture. Cybersecurity regulations impose operational standards. Investment screening reflects concerns about ownership and control.

For nations like India, this strategy balances openness with autonomy. Global hyperscalers bring capital and expertise. Domestic operators provide resilience and local accountability. Policy must enable both without surrendering strategic leverage.

The future is not isolation, but negotiated interdependence.

In federal systems, decentralization creates competition. States vie for investment by offering incentives, infrastructure readiness, and regulatory clarity. When aligned with national goals, this competition accelerates growth. When misaligned, it fragments outcomes.

The most successful regions treat data centers not as one-off projects, but as part of a broader digital ecosystem. They align energy planning, urban development, skilling initiatives, and industrial policy. Without it, even generous incentives fail to deliver durable infrastructure.

Investors and the Long View of Capital

If governments shape the rules, investors shape the scale.

The future of data centers depends increasingly on patient capital. These are assets with long lifespans, high upfront costs, and steady returns rather than explosive growth. Pension funds, sovereign wealth funds, infrastructure funds, and long-term institutional investors are therefore becoming central players.

This shift changes incentives.

Short-term optimization gives way to durability. Energy efficiency matters because operating costs persist for decades. Community acceptance matters because disruption threatens continuity. Regulatory compliance matters because penalties compound over time.

Capital begins to favor predictability over speed.

This coincides, in most instances, with the interest of the people. Sustainable design, workforce development, and resilient infrastructure would be encouraged by long-horizon investors more easily. They rely on stability and not volatility in relation to their returns.

Hyperscalers and Market Power

Global cloud providers occupy a unique position in this landscape.

Their scale enables unmatched efficiency, but it also concentrates power. Their infrastructure decisions influence energy markets, labor demand, and even regional development patterns. Governments court them, yet remain wary of dependency.

This tension defines much of the future policy debate.

How much concentration is acceptable? When does efficiency become fragility? How should public interest be protected without stifling innovation?

There are no simple answers. But transparency, competition policy, and interoperability standards will play growing roles. Policy must therefore be designed with foresight, not reaction.

Communities as Stakeholders, Not Bystanders

Local communities will become active stakeholders in influencing data centers' future.

With the increased size and number of facilities, their effect is felt. Land use changes. Power demand increases. Water usage draws attention. Employment patterns shift.

Communities ask legitimate questions. What do we gain? What do we lose? How are decisions made?

Projects that treat communities as obstacles often face resistance, delays, or legal challenges. Those who engage early, invest locally, and communicate clearly build trust.

This trust becomes a competitive advantage.

Data centers that integrate community benefits, local hiring, renewable projects, heat reuse, and digital education are more likely to receive approvals and social license. The future of data centers will be shaped as much by consent as by capacity.

Regulation in an Era of Rapid Change

Regulation faces a paradox.

Move too slowly, and infrastructure outpaces oversight. Move too quickly, and rules harden before technologies mature. The pace of data center evolution makes this balance difficult.

Effective regulation will therefore emphasize principles over prescriptions.

Rather than specifying technologies, regulators define outcomes. Energy efficiency targets. Resilience standards. Data protection obligations. Environmental thresholds.

This flexibility allows innovation while protecting public interest.

It also demands regulatory capability. Governments must understand the systems they regulate. Capacity building within public institutions becomes critical.

At this stage of the data center journey, the decisive factor is no longer technological feasibility. It is institutional alignment.

Countries and regions that align policy, capital, and community interests will build sustainable digital infrastructure. Those who rely solely on incentives or market forces will encounter friction.

The future of data centers is therefore a governance challenge as much as an engineering one.

Inclusion, Sustainability, and the Measure of Real Progress

As the future of data centers comes into clearer view, one question rises above all others. What, ultimately, are these vast systems for?

For years, the answer was framed narrowly. Data centers enabled cloud services. They supported enterprises. They powered platforms. Their value was measured in uptime, efficiency, and return on capital.

That definition is no longer sufficient.

Data centers are now a significant infrastructure to societies, and not only to markets. They have effects that go beyond balance sheets to education, health, government, and day-to-day living. The future of data centers, therefore, should be evaluated using wider parameters.

Not just how fast they grow, but how well they serve.

Digital Infrastructure as a Social Multiplier

Infrastructure amplifies whatever it supports.

When roads are built, commerce follows. When electricity reaches households, productivity rises. When broadband spreads, access to information expands. Data centers operate at a similar multiplier scale, but with a deeper reach.

They amplify knowledge, coordination, and decision-making.

When digital infrastructure is reliable and proximate, schools can adopt blended learning. Hospitals can integrate diagnostics and telemedicine. Small businesses can access global markets. Governments can deliver services more efficiently.

These outcomes are not automatic. They depend on design choices.

Data centers that prioritizes only the most profitable workloads may reinforce inequality. Those that support public platforms, regional services, and affordable access widen opportunity.

The future, therefore, demands intentionality.

Inclusion Beyond Connectivity

For much of the last decade, digital inclusion was defined as connectivity. Getting people online was the goal. That phase, while incomplete, is no longer enough.

Meaningful inclusion requires performance, reliability, and trust.

A slow or unreliable connection limits participation. Services that fail during peak demand erode confidence. Systems that mishandle data create fear.

Data centers influence all three.

By bringing computing power closer to users, edge infrastructure improves experience quality. By stabilizing platforms, hyperscale resilience supports continuity. By enforcing strong governance, facilities protect trust.

Inclusion is therefore infrastructural as much as social.

Start-ups, SMEs, and Local Innovation

One of the most tangible ways data centers drive inclusive growth is by lowering barriers for innovation.

When computation is affordable and accessible, start-ups can experiment without prohibitive capital costs. SMEs can digitise operations. Entrepreneurs can build services for niche markets.

This dynamic has reshaped economies globally.

Yet it is sensitive to pricing, latency, and reliability. An infrastructure that favours only large clients concentrates the advantage. Infrastructure that supports flexible, scalable access disperses it.

Future data centers will increasingly differentiate themselves not only by capacity, but by ecosystem support. Developer programs, local partnerships, and integration with public platforms extend value beyond tenancy. Data centers, by shaping the cost of experimentation, shape the innovation landscape.

Sustainability as Intergenerational Responsibility

Environmental sustainability is often framed as risk mitigation or compliance. In reality, it is a question of responsibility across time.

Data centers built today will operate for decades. Their energy choices affect emissions trajectories. Their water use affects local ecosystems. Their materials footprint influences resource extraction.

Future-oriented operators recognize this temporal dimension.

Green-first design is not merely about meeting current standards. It is about anticipating stricter ones. Facilities that rely on fragile energy assumptions may become stranded assets. Those integrated with renewable ecosystems remain viable.

Sustainability, therefore, aligns with long-term resilience.

The future of data centers must respect this continuity.

The Quiet Power of Reliability

Perhaps the greatest contribution of future data centers will remain invisible.

When systems work reliably, societies forget about them. Payments clear, Records persist, Communications flow, and Services remain available during crises.

This quiet reliability underpins trust.

During natural disasters, resilient infrastructure supports emergency response. During economic shocks, it maintains continuity. During social transitions, it provides stability.

In this sense, data centers act as shock absorbers for modern life.

Their value emerges most clearly when conditions are worst.

The future of data centers is not defined by a single technology or trend. It is shaped by a convergence of intelligence, energy, governance, and social expectation.

Data centers are becoming mirrors of the societies they support. Their design reflects collective priorities. Their failures expose collective neglect.

The question is no longer whether data centers will grow. They will.

The question is what kind of growth they will embody.

The Infrastructure We Choose to Become

"We shape our buildings, and afterward our buildings shape us." – Winston Churchill.

Towards the conclusion of this long arc, it is made clear that the future of data centers cannot be simplified to technology projections or capacity forecasts. It does not just concern the quicker processors, clever cooling, or the more environmentally conscious power contracts. It is about choice.

Any physical expression of intent is in every data center. The place of its construction, its source of power, the people that run it, and those that it serves are some of the choices that evoke values as much as its financial impact. Such decisions, with time, give rise to systems that shape the functioning of societies in a silent way.

The future of data centers, therefore, is not something that happens to us. It is something we enact.

From Supporting Role to Systemic Backbone

Data centers have evolved from supporting roles, like hosting websites and storing records, to becoming essential infrastructures that underpin critical systems in finance, healthcare, public administration, transportation, energy management, and communications. This shift brings significant responsibilities, as outages or breaches can impact entire sectors and public trust.

The future data center must be treated as civic infrastructure, regardless of ownership, as private systems that are essential carry public obligations. Success is now defined by how well these centers facilitate economic growth, align with policies, ensure sustainability, and maintain reliability.

The future of data centers lies at the intersection of these factors, recognizing that there's no one-size-fits-all model. Different regions will adopt varying strategies, some emphasizing efficiency, others decentralization or renewable integration. Ultimately, coherence among technology, policy, energy, and societal expectations will determine long-term success. Systems that optimize one area while neglecting others will face challenges.

Data Centers as Engines of Inclusive Growth

At their best, data centers expand opportunity.

Small firms gain the ability to reach global markets, creating new opportunities for growth. Governments can provide services more efficiently, ensuring that resources are utilized effectively. Furthermore, essential sectors such as education, healthcare, and financial inclusion receive much-needed support. This progress also helps lower the costs associated with coordinating efforts across large populations, streamlining processes, and enhancing overall efficiency.

These benefits are not automatic. They depend on access, affordability, and proximity.

Infrastructure that concentrates value while externalizing cost undermines inclusion. Infrastructure that distributes capability while respecting limits supports it.

The future data center must therefore be evaluated not only by who builds it, but by who benefits from it. This perspective reframes investment decisions.

Sustainability as a Condition, not a Constraint

For much of the industry's history, sustainability was framed as a constraint on growth. Energy limits, water scarcity, and environmental regulation were obstacles to overcome.

In the future, sustainability will become a condition of legitimacy.

Facilities that cannot align with energy transitions will face resistance. Those that strain local resources without reciprocal benefit will lose social license. Those who ignore climate risk will suffer disruption.

Conversely, data centers that integrate with renewable systems, support grid stability, reuse waste heat, and reduce environmental impact will gain a long-term advantage.

Sustainability ceases to be a trade-off. It becomes a prerequisite.

Trust as the Ultimate Currency

As digital dependence deepens, trust becomes the most valuable asset data centers possess.

Trust that data will be protected. Trust that systems will remain available. Trust that infrastructure will not become a tool of exclusion or control.

This trust is fragile. It is built slowly and lost quickly.

Future data centers must therefore prioritize transparency, accountability, and governance alongside performance. They must be designed not only to function, but to be trusted.

This trust extends beyond users to regulators, communities, and future generations.

The Long View of Infrastructure

Perhaps the most important insight is temporal.

Data centers have been built for decades. Their consequences extend far beyond current market cycles or political terms. Decisions made today will shape digital capacity, energy demand, and social outcomes long into the future.

This long view demands humility.

No planner can predict all future needs. No operator can foresee every risk. The task, therefore, is not to build perfect systems, but adaptable ones.

Infrastructure that can evolve, respond, and learn will outlast infrastructure that merely optimizes.

A Quiet Conclusion

The future of data centers will not announce itself with a single breakthrough or turning point. It will emerge quietly, through thousands of design choices, policy decisions, and operational practices.

Most people will never see these facilities. Many will never think about them. That invisibility is part of their success. But their influence will be everywhere.

In the speed of a transaction. In the reliability of a public service. In the resilience of a city during a crisis. In the reach of opportunity across regions. Data centers are not just places where data lives. They are places where futures are negotiated.

This book began by explaining what data centers are and how they evolved. It traced their economic impact, policy challenges, energy demands, risks, and human dimensions. It ends here, with a recognition.

Data centers are no longer quiet backrooms of the digital world. They are its foundations.

The question that remains is not whether we will build them, but how wisely we will do so.

Because in building them, we are also building the conditions under which societies will live, work, and govern themselves in the decades to come.

"The future is built one structure at a time." – A modern axiom.

Verses Kindler Publication

Reach us through our website -

https://www.verseskindlerpublication.com/

For more information visit our

Instagram or Facebook page.

www.ingramcontent.com/pod-product-compliance
Lightning Source LLC
LaVergne TN
LVHW010714061126
851857LV00014B/1186
* 9 7 8 8 1 6 9 2 4 8 0 8 2 *